Spinel and Inverse Spinel Ferrites

This comprehensive book explores spinel and inverse spinel ferrites, focusing on their synthesis methods, structural characteristics, magnetic properties, and diverse applications. It offers a valuable resource for understanding how these materials are transforming fields such as electronics, energy conversion, sensing, biomedicine, agriculture, and environmental management. The book provides practical insights into synthesis methods, fabrication techniques, and the scale-up processes required to move these materials toward commercial applications. With a focus on recent advancements such as nanoscale engineering and surface modifications, the book offers readers insights into the commercial and practical potential of these materials across various industries.

- Delivers practical guidance on the synthesis, fabrication, and scale-up of these materials, addressing their commercialization prospects.
- Examines the role of spinel and inverse spinel ferrites in magnetic resonance imaging (MRI), exploring their applications in medical diagnostics and treatment.
- Discusses their effectiveness in electromagnetic interference (EMI) shielding, emphasizing the importance of ferrites in electronic and telecommunication devices.
- Provides insights into the application of ferrites as sensors, with a focus on their use in gas sensing, biosensing, and other diagnostic tools.
- Highlights photocatalytic activity and environmental remediation, showcasing how these materials help in pollution control, water purification, and sustainable energy solutions.

This reference book is for students, researchers, and professionals in physics, materials science, and engineering who wish to deepen their understanding of spinel and inverse spinel ferrites and their interdisciplinary applications.

Spinel and Inverse Spinel Ferrites

Theoretical and Technological Advances

Edited by
Naveen Kumar, Nupur Aggarwal, and
Anjana Sharma

CRC Press
Taylor & Francis Group
Boca Raton London New York

CRC Press is an imprint of the
Taylor & Francis Group, an **informa** business
A CHAPMAN & HALL BOOK

Designed cover image: ShutterStock

First edition published 2025
by CRC Press
2385 NW Executive Center Drive, Suite 320, Boca Raton FL 33431

and by CRC Press
4 Park Square, Milton Park, Abingdon, Oxon, OX14 4RN

CRC Press is an imprint of Taylor & Francis Group, LLC

ISBN: 978-1-032-66455-2 (hbk)
ISBN: 978-1-032-66563-4 (pbk)
ISBN: 978-1-032-66565-8 (ebk)

DOI: 10.1201/9781032665658

Typeset in Times
by Deanta Global Publishing Services, Chennai, India

Contents

Preface

Spinel and inverse spinel ferrites have become focal points of both fundamental research and practical applications owing to their diverse and innovative physiochemical properties. Particularly in their nanostructured forms, these magnetic materials exhibit a range of morphologies that have captured the attention of researchers across various fields. Their distinct spinel structure enables the creation of a vast array of ferrite compositions, making them highly desirable for applications in advanced electronics, microwave technologies, and biomedical fields. Despite the abundance of literature on spinel nanoferrites and their industrial uses, there exists a significant gap in comprehensive resources that consolidate their structures, properties, and advanced applications, leaving much to be desired.

This book endeavors to fill this gap by offering a comprehensive exploration of the complexity surrounding spinel and inverse spinel nanoferrites. It meticulously details their synthesis, physicochemical properties, and potential applications across multiple high-tech fields. Catering to a diverse audience ranging from beginners to seasoned specialists in academic and industrial sectors, the chapters provide a thorough understanding of spinel and inverse spinel ferrites. Beginning with an introduction to these materials and their unique characteristics, the book progresses to delve into detailed discussions on synthesis and characterization, setting the stage for exploring their diverse applications.

The book further delves into the advanced applications of spinel and inverse spinel ferrites, such as their utilization in magnetic resonance imaging (MRI), electromagnetic shielding, drug delivery systems, agriculture, gas sensors, etc., showcasing their relevance in various cutting-edge technologies. It also explores their potential in microwave absorption, energy storage, and agricultural enhancements, demonstrating their versatility and impact across multiple industries. Additionally, the book addresses current challenges and future prospects, providing valuable insights into ongoing research and potential advancements in the field. This comprehensive resource aims to equip researchers, students, and professionals in the field of materials science and engineering with a holistic understanding of spinel and inverse spinel ferrites, making it an indispensable reference for those seeking to explore their properties and applications.

Editors

Naveen Kumar is Associate Professor in the Department of Physics at Galgotias Multidisciplinary Research and Development Cell (G-MRDC), Galgotias University, India. He formerly worked as Associate Professor at in the department of University Centre for Research & Development at Chandigarh University, India. He works on interdisciplinary research fields involving very recent and technologically significant strands: the synthesis of multiferroics for dielectric and ferroelectric studies. He has been working in the field of magnetic ferrites such as $BiFeO_3$, Fe_3O_4, and $Mg_{0.05}Zn_{0.05}Fe_2O_4$ for the past eight years and has published more than 100 research articles in peer-reviewed SCI/SCOPUS-indexed journals and several research papers in international conferences. He is also an editor of multiple journals published by reputed publishers.

Nupur Aggarwal is Associate Professor at Chandigarh University, India. She spearheads interdisciplinary research in materials science. Her work focuses on the synthesis of metal oxide nanoparticles, targeting applications in optoelectronics and spintronics. Over the past nine years, Dr Aggarwal has delved deeply into the field of dilute magnetic semiconductors, striving to unlock new potentials in electronic and spintronic devices by enhancing their magnetic and electrical properties. With more than 50 research publications in reputed SCI and SCOPUS-indexed journals, she plays a vital role as a reviewer and editor for esteemed journals, where she helps shape the direction of future research in nanotechnology.

Anjana Sharma is Associate Professor at Chandigarh University, India. She is known for her work in the field of interdisciplinary materials science. Dr Sharma specializes in synthesizing ferrites for dielectric studies, focusing on the effects of transition metal doping on the physicochemical properties, cation distribution, and metal site preferences in spinel and inverse spinel nanoferrites. She has published more than 20 research articles in refereed journals, showcasing her findings on the synthesis, characterization, and application of nano-ferrites. Additionally, she has presented over 15 articles at international conferences, sharing her insights with the global scientific community and fostering collaborations with researchers from around the world.

Contributors

Nupur Aggarwal
Chandigarh University
Mohali, Punjab, India

Mohammad Abass Ahanger
Northwest A&F University
Shaanxi, China

Parvaiz Ahmad
Department of Botany and Microbiology
King Saud University
Riyadh, Saudi Arabia

Pravej Alam
Prince Sattam bin Abdul Aziz University
Alkharj, Saudi Arabia

Sandeep Arya
University of Jammu
Jammu and Kashmir, India

Kanchan Bala
Government Mohindra College
Patiala, India

Harpreet Kaur Channi
Chandigarh University
Mohali, Punjab, India

Ankush Chauhan
Chettinad Hospital and Research Institute
Tamil Nadu, India

Fateh Singh Gill
Graphic Era (Deemed to be University)
Dehradun, Uttarakhand, India

Anu Kapoor
Chandigarh University
Mohali, Punjab, India

Kamaldeep Kaur
Chandigarh University
Mohali, Punjab, India

Navjeet Kaur
Chandigarh University
Mohali, Punjab, India

Sandeep Kaur
Chandigarh University
Mohali, Punjab, India

Cengiz Kaya
Plant Nutrition and Physiology
Harran Üniversitesi
Turkey

Chandra Kumar
Engineering and Technology
Universidad Mayor
Santiago, Chile

Manjeet Kumar
Multidiscpilinary Core Institute for Future Energies
Incheon National University
Yeonsu-gu, Incheon, Republic of China

Naveen Kumar
Chandigarh University
Mohali, Punjab, India
and
Galgotias University
Greater Noida, Uttar Pradesh, India

Pradeep Kumar
University Institute of Technology
Himachal Pradesh University
Himachal Pradesh, India

Pulkit Kumar
Chandigarh University
Mohali, Punjab, India

Shubham Kumar Patial
Academy of Scientific and Innovative Research (AcSIR)
Ghaziabad, Uttar Pradesh, India
and
Materials Science and Sensor Applications
CSIR-Central Scientific Instruments Organisation
Chandigarh, India

Madan Lal
Graphic Era (Deemed to be University)
Dehradun, Uttarakhand, India

Manju
Central University of Kashmir
Ganderbal, Jammu and Kashmir, India

Nancy
Department of Physics
Amity University Haryana
Gurugram, Haryana, India

Aditya Nath Mishra
Wigner Research Centre for Physics
Budapest, Hungary

Akhil Nigam
Chandigarh University
Mohali, Punjab, India

Anuj Sharma
CSIR-Central Scientific Instruments Organisation
Chandigarh, India

Deepika Sharma
IEC University
Solan, Himachal Pradesh, India

Navdeep Sharma
Abhilashi University
Chailchowk, Himachal Pradesh, India

Fateh Singh Gill
Graphic Era (Deemed to be University)
Dehradun, Uttarakhand, India

Suman Singh
Academy of Scientific and Innovative Research (AcSIR)
Ghaziabad, Uttar Pradesh, India

Atul Thakur
Amity University
Haryana, India

Preeti Thakur
Amity University
Haryana, India

Priyanka Thakur
Eternal University
Himachal Pradesh, India

Ritesh Verma
Amity University
Haryana, India

Neelam Vishwakarma
Academy of Scientific and Innovative Research (AcSIR)
Ghaziabad, Uttar Pradesh, India

Leonard Wijaya
National Research and Innovation Agency
Indonesia

1 Introduction to Spinel and Inverse Spinel Ferrites

Manju and Kamaldeep Kaur

1.1 INTRODUCTION

The tendency of a material to respond to the applied magnetic field resulted in a new era of science, which classifies materials based on their extent of reaction to the external magnetic field. On a scale of magnetic susceptibility ranging from low to high, magnetic materials are classified as paramagnetic, diamagnetic, ferromagnetic, antiferromagnetic, and ferrimagnetic. Among ferrimagnets, a class of materials is called ferrites, which hold significant consideration due to their tendency to generate different geometric configurations through varying arrangements of metal cations and oxygen anions [1, 2]. Ferrites exhibit high specific resistance and flexibility, along with experimental ease of preparation, giving a good balance between price and performance. Owing to such a splendid set of properties, ferrites find their application in numerous fields, such as memory chips, antenna rods, activators, computer components, drug delivery systems, transducers, permanent magnets, integrated circuitry, millimeter and microwave devices, spintronics, etc. [3, 4].

Ferrites are classified based on their structure as garnets, hexaferrites, orthoferrites, and spinel ferrites. Among all these classes, Fe and O are present as cations and anions with different arrangements with other atoms. A brief description of these classes is as below:

a) Garnets: The general formula for garnet-based ferrites is $R_3Fe_5O_{12}$, where R stands for a rare earth element. This structure has tetrahedral, octahedral, and dodecahedral crystallographic sites [5].
b) Hexaferrites: The general formula for hexaferrites is $MFe_{12}O_{19}$, where M is generally Ba, Sr, Ca, or Pb. These have complex crystal structures, with two crystallographic sites for the cations, i.e., tetrahedral and octahedral. These are hard magnets [6].
c) Orthoferrites: This class of materials is a subgroup of the orthorhombic family, with space group *Pbnm*. The general formula for orthoferrites is $RFeO_3$, with R being a rare earth element. These are weak ferromagnets [7].

DOI: 10.1201/9781032665658-1

d) Spinel ferrites: Spinel indicates a cubic crystal structure, with the general formula AB_2O_4, where A and B sites are occupied by divalent and trivalent cations, respectively. Thus, spinel ferrites have a general formula of AFe_2O_4, with A being Ni, Co, Zn, Mg, Mn, Cu, Cd, etc. This crystal system has a closely packed cubic structure, with a semiconducting nature of the system [8].

Among these, spinel ferrites possess splendid magnetic properties such as a high squareness ratio, high permeability, low coercivity, high electric resistivity, high saturation magnetization, large magneto-crystalline anisotropy, high Curie temperature, low eddy current losses, etc. [9]. Moreover, spinel ferrites offer a variety of ways to be synthesized in bulk, nano, or thin film form. Hence, the present chapter is dedicated to highlighting the structure of spinel ferrites, factors affecting the magnetic response, a brief overview of various synthesis techniques, and examples of extensively studied spinel ferrite systems.

1.2 STRUCTURE

As introduced above, spinel ferrites belong to the face-centered cubic family, with the general formula AFe_2O_4, where A is any divalent cation, referred to as A(II). Hence, there is one divalent A(II) cation and one trivalent Fe(III) cation, with a total of two crystallographic sites for cations. The A(II) cation is bonded with four oxygen atoms and Fe(III) cation is bonded with six oxygen atoms, forming tetrahedral and octahedral sites, respectively. Each unit cell has 8 formula units with a total of 96 interstices, namely 64 tetrahedral sites, 32 octahedral sites, and 32 sites of oxygen cations, out of which only 8 tetrahedral and 16 octahedral sites are occupied, and the rest are vacant sites. The tetrahedral sites are called network formers, and octahedral sites are called network modifiers [10]. Moreover, depending on the relative occupancy of A(II) and Fe(III) cations on tetrahedral and octahedral sites, the spinel structure is again divided into three categories: normal, inverse, and mixed spinels, as shown in Figure 1.1.

The complete general formula for spinel ferrites is $(A^{2+}{}_{1-x}Fe^{3+}{}_{x})[A^{2+}{}_{x}Fe^{3+}{}_{2-x}]O^{2-}{}_{4}$, where the parentheses stand for the tetrahedral site and square brackets stand for the octahedral site. The x in the subscript stands for the degree of inversion, which relates

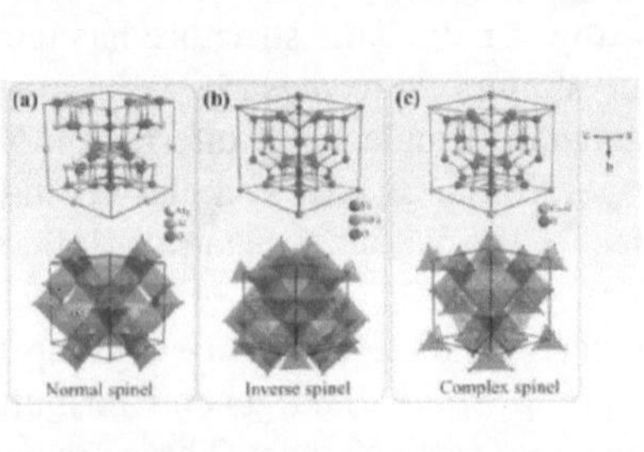

FIGURE 1.1 The normal, inverse, and mixed/complex spinel structures, with the example of (a) $MgAl_2O_4$, (b) $NiFe_2O_4$, and (c) $CuAl_2O_4$. Reprinted with permission from [11].

to the tetrahedral and octahedral occupation by A(II) and Fe(III) cations. In normal spinel ferrites, i.e., for $x = 0$, all the tetrahedral sites are occupied by A(II) cations, and all the octahedral sites are occupied by Fe(III) cations. In this case, A(II) acts as the network former, and Fe(III) acts as the network modifier. In inverse spinels, i.e., $x = 1$, A(II) cations occupy half of the octahedral sites, but all of the tetrahedral sites are occupied by half of the Fe(III) cations. In other words, all A(II) are present on octahedral sites, but Fe(III) are present equally on tetrahedral and octahedral sites in the inverse spinel ferrite structure. Thus, Fe(III) acts as both a network former and a network modifier. In mixed spinels, having a value of x between 1 and 0, a mixture of cation arrangements of normal and inverse systems is observed. Thus, in spinel ferrites, the exchange interaction of valence electrons of tetrahedrally and octahedrally situated cations decides the net magnetic moment of the compound. As per Néel's theory of ferrimagnetism, the magnetic moment of individual sub-lattices sums up to give the total magnetic moment to the ferrite compound. If for normal spinels, A(II) is located at tetrahedral sites and Fe(III) is located at octahedral sites, then antiferromagnetic exchange interactions between A(II) and Fe(III) are the strongest. On the other hand, the ferromagnetic interactions of A(II)–A(II) are weaker than the A(II)–Fe(III) interaction by the order of ten, whereas ferromagnetic interactions of Fe(III)–Fe(III) are the weakest. But, with variation in cation distribution due to variation in the degree of inversion, exchange interactions also get modified, resulting in the modification of net magnetization from the system [12]. As in the case of $CoFe_2O_4$, which is an inverse spinel ferrite, Fe^{3+} occupies all the tetrahedral sites, whereas the octahedral sites are combinedly occupied by Fe^{3+} and Co^{2+}. The magnetic moments of atoms at tetrahedral and octahedral sites have intra-site parallel alignment but inter-site antiparallel alignment. Hence, the net magnetic moment (M_{net}) caused by this alignment is represented in Equation (1.1) [13]:

$$M_{net} = \sum M_{octahderal} - \sum M_{tetrahedral} \tag{1.1}$$

where $M_{octahedral}$ and $M_{tetrahedral}$ represent the magnetic moment of atoms at octahedral and tetrahedral sites, respectively. Hence, it becomes crucial to understand the cation composition as well as their relative arrangement in the spinel ferrites to get a tailor-made response [14].

Various characterization techniques are employed to quantify cationic site occupancy, through the involvement of electronic transitions of metal cations or nuclear transitions of Fe centers in spinel ferrites. Thus, techniques like X-ray spectroscopies, Raman spectroscopy, neutron diffraction, Mössbauer spectroscopy, etc., give precise insights into the coordination sphere of metal ions and the symmetry of the system [15]. For spinel ferrites, the variation in the oxidation state of Fe (Fe^{2+}, Fe^{3+}, or Fe^{4+}) causes variation in s-electron density, which can be judged by either X-ray absorption near edge structure (XANES) analysis [16], X-ray photoemission (XPS) spectroscopy [17], or Mössbauer spectroscopy [18–20]. Nakashima et al. [15] employed the first principles method to examine the cation disordering in $ZnFe_2O_4$ through XANES analysis. The samples were synthesized by various techniques to alter the cation distribution, i.e., the thin films of $ZnFe_2O_4$ were prepared by sputtering. These films were annealed at different temperatures (350, 400, 500, and 800°C), and bulk samples were synthesized by solid-state reaction, to investigate the systematic emergence of cation distribution in correlation with its magnetic response. To dig into the evidence for cation distribution, experimental XANES spectra at Zn *K*-edge were compared with simulated spectra by keeping Zn^{2+} at the A site of normal spinel (Zn_A^{2+}, green), the A site of disordered spinel (Zn_A^{2+}, red), and the B site of disordered spinel (Zn_B^{2+}, blue), as shown in Figure 1.2(a). The comparison of experimental and theoretical spectra revealed that annealing results in the distribution of Zn^{2+} in the normal spinel structure, as is present in bulk samples. A similar treatment was performed for the Fe *K*-edge by keeping Fe^{3+} at the B site of the normal spinel structure (Fe_B^{3+}, blue line) and Fe^{3+} at the A site in the disordered spinel structure (Fe_A^{3+}, red line), as shown in Figure 1.2(b). It was observed from the comparison that Fe^{3+} tends to occupy the B site of the spinel structure, having a close resemblance in the distribution of annealed thin film and bulk specimen. However, the presence of an intense pre-edge feature (A in bulk and F in as-deposited) indicates some A site occupancy of Fe^{3+} ions. The XANES analysis was combined with a temperature-dependent DC magnetization study at zero-field cooled and field-cooled conditions at 50 Oe, and it was noted that with the increase in annealing, the super-exchange interactions among A and B site Fe^{3+} ions weakened with a subsequent increase in super-exchange interactions among B site Fe^{3+} ions, causing the dominant antiferromagnetic behavior. But with an increase in disorder, the ferromagnetic behavior dominated due to super-exchange interactions among A and B site Fe^{3+} ions.

On the other hand, Sumalatha et al. [18] employed detailed Raman and Mössbauer spectroscopic studies to correlate the cation occupancies with the observed magnetic properties of the Ga-substituted $CoFe_2O_4$ system. Through low temperature and high magnetic fields, the separate contributions from atoms residing at tetrahedral and octahedral sites were obtained, as shown in Figure 1.3. The quantitative analysis

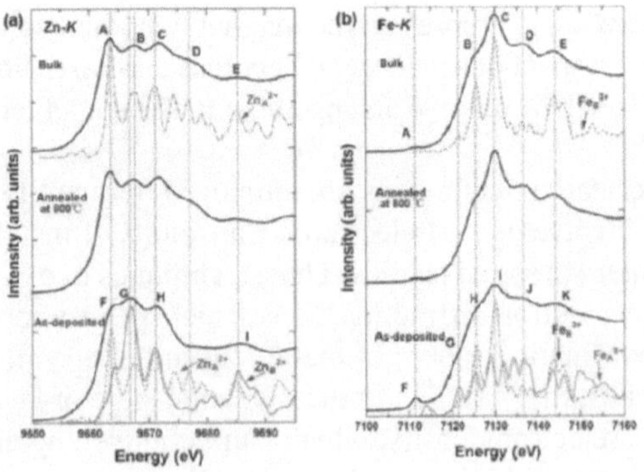

FIGURE 1.2 XANES spectra of experimental (black solid) and theoretical (colored) for bulk, as-deposited, and annealed thin film of $ZnFe_2O_4$ for (a) Zn *K*-edge and (b) Fe *K*-edge. The reader is advised to refer to the text for further information about the color coding of simulated XANES. Reprinted with permission from [15].

of center shift, line width, effective field, fractional area, quadrupole shift, and A_{23} values (ratio of the area of the second and third lines of Mössbauer sextet) helped in understanding the fact that the inclusion of Ga into the system results in a transformation into a mixed spinel structure. The spectra, in correlation with magnetic measurements, indicated the presence of ferrimagnetic ordering in the system. It was also concluded from the combined Raman analysis that Ga substitution forced Fe^{3+} to reside at octahedral sites.

Similarly, instead of the X-ray diffraction study, neutron diffraction is better suited to quantify the cationic distribution in spinel ferrites. The former deals with X-ray scattering from electron density around the metal atoms, which suffers from the indistinguishability of metal atoms having similar electron density. Conversely, neutrons scatter from the nuclei of atoms and give a response depending on the neutron scattering length of the respective element. Hence, researchers rely on the detailed analysis of neutron diffraction data to find cation distribution or inversion degree in spinel ferrites. Wang et al. [21] explored complex spinel ferrites, termed as high entropy oxides, composed of various combinations of Mg^{2+}, Fe^{2+}, Co^{2+}, Ni^{2+}, Cu^{2+}, Zn^{2+}, and Mn^{2+} on the A site of spinel ferrite system. In such cases, detailed structure analysis using Rietveld refinement of neutron diffraction data and neutron pair distribution function (Figure 1.4) helped in diving into the cationic arrangement of these high-entropy spinel ferrites. The structure modeling was done through a combination of normal, inverse, and random arrangement structure models along with a ferrimagnetic ordering model, to incorporate possible site preferences of cations in correlation with their magnetic alignments. The study provided a guide for deciphering the cationic arrangement of spinel ferrites with a judicial choice of structure models.

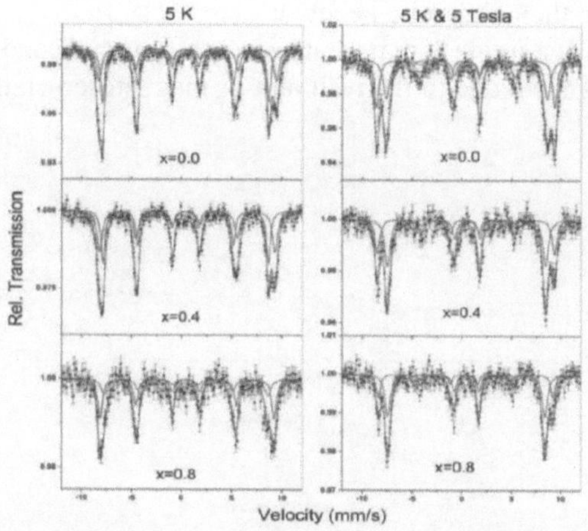

FIGURE 1.3 ^{57}Fe Mössbauer spectra of $CoFe_{2-x}Ga_xO_4$, with the external magnetic field applied parallel to gamma rays. Reprinted with permission [18].

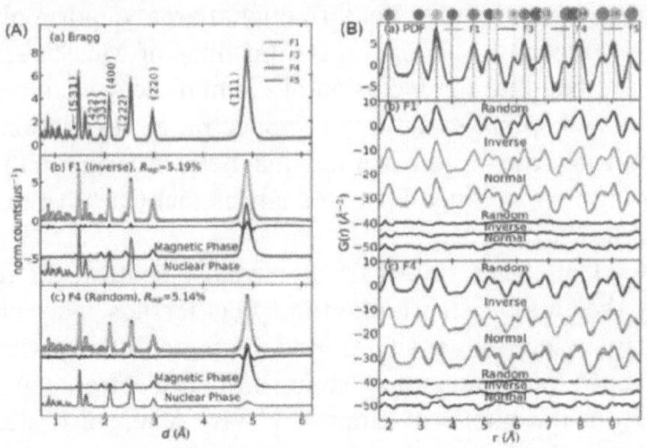

FIGURE 1.4 (A) Represents (a) neutron diffraction data for all samples, (b) the Rietveld refined neutron diffraction patterns for F1 for the inverse spinel model with the ferrimagnetic spin order model, and (c) the Rietveld refined neutron diffraction pattern for F4 for random solution spinel with the ferrimagnetic spin order model. On the other side, (B) represents the pair distribution function (PDF) data for (a) all the samples and (b) and (c) showing fitted data with the indicated structure model. Here, F1, F3, F4, and F5 represent $(Mg_{0.2}Fe_{0.2}Co_{0.2}Ni_{0.2}Cu_{0.2})Fe_2O_4$, $(Mg_{0.2}Co_{0.2}Ni_{0.2}Cu_{0.2}Zn_{0.2})Fe_2O_4$, $(Mg_{0.2}Mn_{0.2}Co_{0.2}Ni_{0.2}Cu_{0.2})Fe_2O_4$, and $(Mn_{0.2}Fe_{0.2}Co_{0.2}Ni_{0.2}Cu_{0.2})Fe_2O_4$, respectively. Reprinted with permission from [21].

To study the magnetic response, it is required to introduce the readers to some of the important parameters, understanding which is crucial to comprehend the magnetic behavior of spinel ferrites. These are diagrammatically depicted with the help of the hysteresis loop shown in Figure 1.5 and are briefly discussed.

1. The size of spinel ferrites is the primary and most important parameter in characterizing their magnetic behavior.
2. The existence of a single domain state is another crucial condition. The magnetic domains form due to the resultant of magneto-crystalline anisotropy

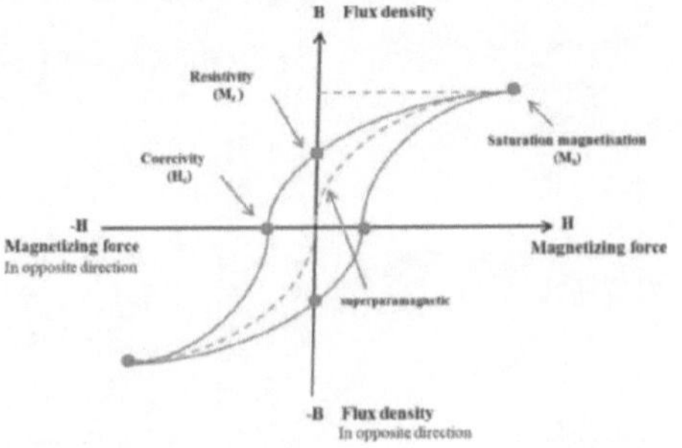

FIGURE 1.5 Hysteresis loop, with important magnetic parameters labeled on it. Reprinted with permission from [22].

and exchange energy, which tend to unwind the walls of magnetic domains, both acting in opposite directions. Thus, at a critical size, the formation of a single domain state is favorable in spinel ferrites, in which it behaves as one giant spin state, even below the Curie temperature.

3. The parameter governing the maximum magnetization allowed at a given temperature is called saturation magnetization.
4. The relaxation time is the parameter corresponding to the flip between two equilibrium states. Thus, the relative comparison of measurement time with relaxation time indicates the possibilities of spinel ferrites being in fluctuated or blocked spin states.
5. The parameter coercivity is also important to understand the surface disorder and surface anisotropy of spinel ferrites. It is obtained from the hysteresis loop.

1.3 FACTORS AFFECTING THE PROPERTIES OF SPINEL FERRITES

The magnetic behavior of spinel ferrites is inherently controlled by various factors, which are briefly discussed.

1.3.1 Size Distribution

The magnetic response of spinel ferrites is significantly affected by particle size. On the transition from bulk to the nano-regime, each system possesses a critical size range below which it loses its paramagnetic behavior under the application of an external magnetic field and becomes superparamagnetic [23]. Each nanoparticle consists of thousands of crystal units, with each crystal unit behaving like a single magnetic domain having a magnetic dipole moment greater than the resultant dipole moment of unpaired electrons present in the crystal. Thus, in the absence of a magnetic field, the net dipole moment is zero due to the thermal randomization of magnetic domains. However, upon application of an external magnetic field, the magnetic domains of nanoferrites align to give a net magnetic moment in the direction of the field. Due to superparamagnetic behavior, nano spinel ferrites do not aggregate magnetic dipoles in any way, which is attributed to strongly negative magnetic interactions among cations at tetrahedral and octahedral sites and negligible interactions at the same sublattice (tetrahedral–tetrahedral or octahedral–octahedral). Hence, below the critical size in the nano-range, the magnetic interactions are solely controlled by thermal effects [24]. At blocking temperature (T_b), the nano-ranged spinel ferrites exhibit superparamagnetic behavior, which is given by the following relation:

$$T_b = KV/a\ k_B \tag{1.2}$$

where K, V, and k_B are the magnetic anisotropy constant, the volume of nano spinel ferrites, and the Boltzmann constant, respectively. The "a" contains information on measurement time (T_m) and relaxation time (T_0), given as $a = \ln(T_m/T_0)$. In general, for $T_m = 100$ s, $T_0 = 10^{-12}$ s, the usual value of a is taken as 25. So, for temperatures less than T_b, the magnetization is proportional to temperature, as thermal energy helps in the alignment of magnetic domains with the applied magnetic field. But, for temperatures greater than T_b, thermal energy perturbs the domain orientation due to randomness, thereby decreasing the magnetization.

The physical nature, such as shape, size, and surface properties of spinel ferrite particles, greatly affects the saturation magnetization, coercivity, and residual magnetization. It is observed that size dictates the saturation magnetization and residual magnetization, whereas coercivity and T_b are controlled by both the shape and size of spinel ferrites. Moreover, some ferrite nanoparticles undergo domain transition from magnetic multidomain to magnetic single-domain state at critical size, at room temperature. The dependency on size and magnetization is self-explanatory. The decrease in size results in the enhancement of surface ions, thereby increasing surface energy that makes it difficult for spin alignment of dipoles. Along with this, the decrease in coordination of surface ions subsequently decreases the saturation magnetization, coercivity, and residual magnetization [25]. Hence, it is crucial to control the size, shape, and crystallinity of spinel ferrites to exploit their magnetic properties for various applications.

1.3.2 Spin Canting Effect

The decrease in size of spinel ferrites results in more surface atoms with broken symmetry, hence resulting in low saturation magnetization. This phenomenon is termed as the core–shell model, where the core consists of normal spin arrangement and the shell has disordered spin arrangement. Thus, the orientation of disordered spin with respect to the ordered shell is called the spin canting angle, which depends on the concentration of disordered spin atoms and dangling bonds on the surface [26]. A schematic of this phenomenon is presented in Figure 1.6.

1.3.3 Presence of Defects

Among various interactions giving rise to the magnetic behavior of spinel ferrites, the vacancies or defects result in a geometrically frustrated spinel system, which competes with existing interactions to influence the magnetic response of the system. As for $ZnFe_2O_4$, the effect of oxygen vacancy on cation ordering and hence on magnetic properties is extensively investigated. The presence of oxygen vacancy interrupts super-exchange interactions and paves the way for cation rearrangement. The oxygen vacancy reduces Zn-O strength and strengthens Fe-O bonds, which subsequently alter the super-exchange interactions [27]. Torres et al. [28] reported that oxygen vacancies induce the local ferromagnetic coupling among octahedrally situated Fe^{3+} ions at Curie temperatures above room temperature, in a usually antiferromagnetic compound. The presence of vacancy influences the swapping of cationic sites in the spinel structure, as shown in Figure 1.7.

Similarly, Zviagin et al. [29] discussed in detail the variation of oxygen vacancies in $ZnFe_2O_4$ spinel ferrites by annealing the compound in oxygen (low and high partial pressure) and argon atmosphere. For the films deposited in an oxygen-deficient

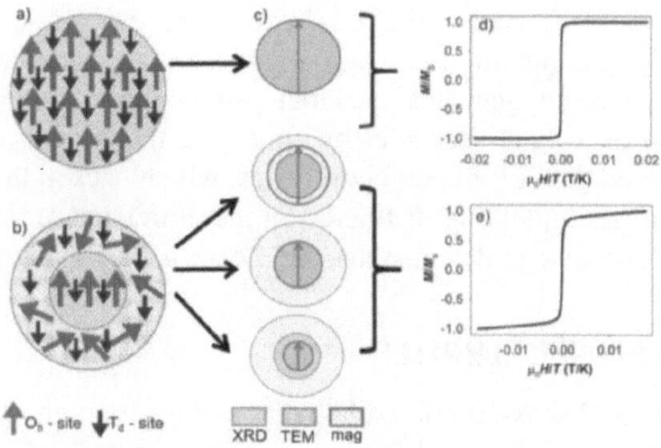

FIGURE 1.6 A schematic of the core–shell structure model with (a) an ideal system and (b) a nanoparticle system. (c–e) Representation of the core-shell model shows the variation of particle size on Langevin curves for ideal and core–shell nanoparticles, indicating the paramagnetic contribution from disordered surface spins. Reprinted with permission from [23].

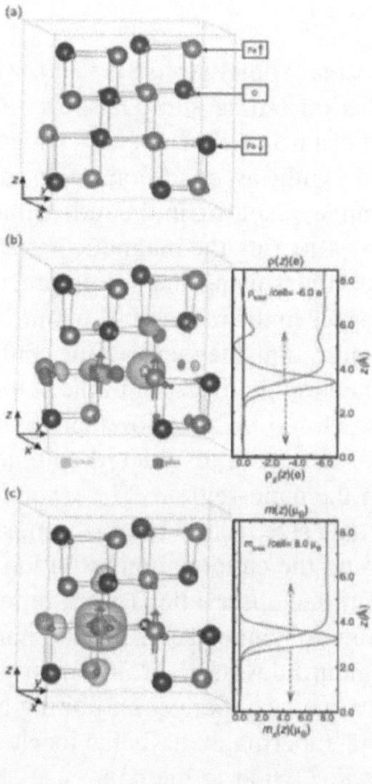

FIGURE 1.7 The ab initio results of the effect of oxygen vacancy on the magnetic structure. (a) Supercell of ideal ZnFe2O4. The oxygen vacancy in the rest of the structures is shown as a white cross. The isosurfaces in the difference in (b) charge density and (c) magnetization around oxygen vacancy are depicted on the left-hand side, and the right-hand side shows graphs of respective quantities integrated over the $x - y$ plane (blue line) and additional integration over z-direction (red line). Reprinted with permission from [28].

environment, Fe^{3+} was observed at the tetrahedral site, giving rise to high magnetization. However, upon annealing in an argon environment, tetrahedral situated Fe^{3+} decreases and enhanced oxygen vacancies increase the magnetization in the system. Lastly, the removal of oxygen vacancies by annealing the system in an oxygen-rich environment migrated the Fe^{3+} to octahedral sites, which backed the local ferrimagnetic ordering. Further annealing at higher temperatures caused recrystallization, with a subsequent decrease in defect states and hence a dip in magnetic response.

1.4 VARIOUS SPINEL FERRITES

Apart from the above-discussed point, chemical composition greatly influences the overall magnetic behavior of spinel ferrites, which is controlled effectively by the synthesis process. Hence, spinel ferrites allow the tuning of their magnetic behavior by controlling the factors responsible for emanating a particular property. Among various spinel ferrite compositions, some of the extensively studied ferrite systems are discussed.

1.4.1 Zinc Ferrite ($ZnFe_2O_4$)

Bulk $ZnFe_2O_4$ exhibits a normal spinel structure, with Zn(II) being on tetrahedral and Fe(III) being on octahedral lattice sites. It shows antiferromagnetism with a Néel temperature of 10 K because of the super-exchange interaction of B–B ions. However, with a change in synthesis conditions, the particle size of $ZnFe_2O_4$ is brought down to the nano-range, resulting in the redistribution of cations on tetrahedral/octahedral sites, thereby affecting the magnetic properties. Upadhyay et al. [30] synthesized $ZnFe_2O_4$ nanoparticles using the citrate precursor route and hydrothermal route, yielding particle sizes in the range of 4–6 nm. The synthesized compound exhibited large values of saturation magnetization, confirming the ferromagnetic ordering, which was attributed to the inversion in the spinel system causing A–O–B superexchange interactions. Moreover, the slow approach to saturation confirmed that the surface of nanoparticles has a disordered spin configuration, owing to the reduction of particle size in the nano-regime. The inversion increases from 40% to 80% with a decrease in particle size from 6 nm to 4 nm, suggesting a crucial role of electrostatic energy in shaping the cationic configuration and subsequent magnetic response of the material. Similar observations were reported by Yao et al. [31], in which thermal decomposition of synthesized $ZnFe_2O_4$ nanoparticles exhibited a 9.8 nm particle size (with a standard deviation of 30%). At this reduced size, $ZnFe_2O_4$ exhibited an inverse spinel structure, thereby exhibiting a saturation magnetization of 65.4 emu g^{-1} at 10 K and supermagnetic behavior at room temperature, with a blocking temperature of 68 K. Owing to the nano-size, the system exhibited small particle anisotropy due to surface spin canting, showing a coercive field of 102 Oe in the blocked state.

1.4.2 Cobalt Ferrite ($CoFe_2O_4$)

$CoFe_2O_4$ belongs to the family of hard magnetic materials, exhibiting an inverse/mixed spinel structure. It possesses a bulk high coercivity of 5.4 kOe and a Curie temperature of 520°C. The highest saturation magnetization of 80 emu g^{-1} at room temperature is reported in this material [32]. Qu et al. [33] reported the effect of reaction temperature on the structure and magnetic properties of $CoFe_2O_4$, synthesized through the coprecipitation technique. The particle size increased from 2.4 nm to 47.4 nm with an increase in reaction temperature from 40°C to 100°C, which further increased the saturation magnetization with a maximum value of 29.5 emu g^{-1} at room temperature. Apart from that, the coercivity also exhibited a size-dependent trend, giving a maximum value of 3267 Oe. Apart from particle size, Zi et al. [34] reported the redistribution of Co^{2+} ions on tetrahedral and octahedral lattice sites when synthesized the same coprecipitation method. For particle sizes ranging between 20 and 30 nm, the system exhibited a saturation magnetization of 61.77 emu g^{-1} at room temperature. The coercivity of 519 Oe was reported, which is attributed to mono-domain behavior, along with transformation from ferrimagnetic to superparamagnetic state at a Curie temperature of 677 K. The observed changes were ascribed to the Co^{2+} ions redistribution at tetrahedral and octahedral sites.

1.4.3 Iron(III) Oxide (ε-Fe_2O_3)

The unit cell of ε-Fe_2O_3 consists of 24 octahedral and 36 tetrahedral positions, among which only 12 octahedral positions are occupied by Fe^{3+} ions. In the case of ordered distribution, the octahedral-to-tetrahedral ratio is expected to be 3:1, whereas in the disordered case, it is 3:2. The degree of ordering is strongly influenced by particle size and hence controls the magnetic behavior of the system. ε-Fe_2O_3 exhibits high room temperature coercivity of 20 kOe due to its canted antiferromagnetic behavior. The system exhibits an effective magnetic moment per Fe^{3+} as 0.3 μ_B [35]. However, Dubrovskiy et al. [36] presented a detailed study on the effect of particle size on the structure as well as magnetic properties of the ε-Fe_2O_3 system, synthesized by the wet chemical pore filling impregnation method. On varying the iron concentration, the samples were found to possess particle sizes ranging from 6 nm to as large as 25 nm. However, a decrease in particle size invited structure disordering, having a large population of Fe^{3+} on the tetrahedral position in a sample with less iron content. Due to this, the sample exhibited a high effective magnetic moment per Fe^{3+} ion to be 0.76 μ_B. The small-sized particles exhibited dominant surface magnetic anisotropy, giving a paramagnetic subsystem behavior. Hence, the system is likely to exhibit tailor-made magnetic properties upon changing particle size.

1.4.4 Nickel Ferrite ($NiFe_2O_4$)

$NiFe_2O_4$ exhibits an inverse spinel structure in bulk form, but may undergo cation redistribution with a change in particle size or synthesis conditions. The system exhibits ferromagnetic character due to the resultant magnetic moment from antiparallel spins between tetrahedral Fe^{3+} and octahedral Ni^{2+} sites. However, the canting of surface spins may happen due to Ni^{2+} occupancy on the tetrahedral site. George et al. [37] reported the effect of grain size on the structural and magnetic properties of $NiFe_2O_4$ synthesized by the sol–gel method. It was observed that saturation magnetization decreased with a decrement in grain size due to cation redistribution and surface spin states. On the other hand, coercivity exhibited direct proportional behavior with grain size. Similarly, Li et al. [38] studied the detailed effect of synthesis conditions on magnetic parameters. In $NiFe_2O_4$ synthesized through a hydrothermal process, an increase in urea content decreases the particle size from 18.9 nm to 14.1 nm, thereby increasing the saturation magnetization from 13.3 to 21.3 emu g^{-1} at room temperature. The coercivity also decreased from 38.5 to 30.8 Oe. It was further reported that an increase in metal precursor concentration increased both particle size and magnetic properties. However, an increase in pH value reduced the particle size, lowered the coercivity, and increased the saturation magnetization due to the relaxation of strains insmall particles.

1.5 CONCLUSION

In conclusion, the study of spinel ferrites offers a fascinating journey into the realm of magnetic materials with diverse structures and versatile properties. Throughout this chapter, we have examined the multitude of factors influencing the properties of spinel ferrites, emphasizing the importance of compositional variations in tailoring their magnetic, electrical, and mechanical characteristics. As we conclude our discussion, it becomes evident that spinel ferrites hold immense potential for various applications across numerous fields. From their utilization in telecommunications for high-frequency devices to their role in biomedical engineering for targeted drug delivery systems, spinel ferrites continue to captivate researchers and engineers alike with their unique properties and wide-ranging functionalities.

REFERENCES

1. Mitsuo Sugimoto. The past, present, and future of ferrites. *Journal of the American Ceramic Society*, 82(2):269–280, 1999.
2. Raul Valenzuela. Novel applications of ferrites. *Physics Research International*, 1:591839, 2012.
3. Martha Pardavi-Horvath. Microwave applications of soft ferrites. *Journal of Magnetism and Magnetic Materials*, 215:171–183, 2000.
4. Shoyebmohamad F Shaikh, Mohd Ubaidullah, Rajaram S Mane, and Abdullah M Al-Enizi. Types, synthesis methods and applications of ferrites. In *Spinel Ferrite Nanostructures for Energy Storage Devices*, pages 51–82. Elsevier, 2020.
5. EJJ Mallmann, ASB Sombra, JC Goes, and PBA Fechine. Yttrium iron garnet: Properties and applications review. *Solid State Phenomena*, 202:65–96, 2013.

6. C de Julián Fernández, Claudio Sangregorio, Juan de la Figuera, B Belec, D Makovec, and Adrián Quesada. Progress and prospects of hard hexaferrites for permanent magnet applications. *Journal of Physics D: Applied Physics*, 54(15):153001, 2021.
7. Majid Abdellahi, Ali Shokrollah Abhari, and Maryam Bahmanpour. Preparation and characterization of orthoferrite $PrFeO_3$ nanoceramic. *Ceramics International*, 42(4):4637–4641, 2016.
8. Sukhleen Bindra Narang and Kunal Pubby. Nickel spinel ferrites: A review. *Journal of Magnetism and Magnetic Materials*, 519:167163, 2021.
9. Hong Qin, Yangzhuo He, Piao Xu, Danlian Huang, Ziwei Wang, Han Wang, Zixuan Wang, Yin Zhao, Quyang Tian, and Changlin Wang. Spinel ferrites ($MgFe_2O_4$): Synthesis, improvement and catalytic application in environment and energy field. *Advances in Colloid and Interface Science*, 294:102486, 2021.
10. VAM Brabers. Progress in spinel ferrite research. *Handbook of Magnetic Materials*, 8:189–324, 1995.
11. Qing Zhao, Zhenhua Yan, Chengcheng Chen, and Jun Chen. Spinels: Controlled preparation, oxygen reduction/evolution reaction application, and beyond. *Chemical Reviews*, 117(15):10121–10211, 2017.
12. Raghvendra Singh Yadav, Jaromir Havlica, Jiri Masilko, Lukas Kalina, Miroslava Hajdúchová, Vojtěch Enev, Jaromir Wasserbauer, Ivo Kuřitka, and Zuzana Kozakova, Structural, caion distribution and magnetic properties of CoFe2O4 spinel ferrite nanoparticles synthesized using a starch-assisted sol-gel auto-combustion method. *Journal of Superconductivity and Novel Magnetism*, 28:1851–1861, 2015.
13. Xue Zeng, Junwei Zhand, Shimeng Zhu, Xia Deng, Hongbin Ma, Junli Zhang, Qiang Zhang, Peng Li, Desheng Xue, Nigel J. Mellors, Xixiang Zhang and Yong Peng Direct observation of cation distributions of ideal inverse spinel $CoFe_2O_4$ nanofibres and correlated magnetic properties. *Nanoscale*, 9(22):7493–7500, 2017.
14. Mahnaz Amiri, Masoud Salavati-Niasari, and Ahmad Akbari. Magnetic nanocarriers: Evolution of spinel ferrites for medical applications. *Advances in Colloid and Interface Science*, 265:29–44, 2019.
15. Seisuke Nakashima, Koji Fujita, Katsuhisa Tanaka, Kazuyuki Hirao, Tomoyuki Yamamoto, and Isao Tanaka. First-principles XANES simulations of spinel zinc ferrite with a disordered cation distribution. *Physical Review B*, 75(17):174443, 2007.
16. David C Bock, Killian R Tallman, Haoyue Guo, Calvin Quilty, Shan Yan, Paul F Smith, Bingjie Zhang, Diana M Lutz, Alison H McCarthy, Matthew M Huie, et al. (De) lithiation of spinel ferrites fe $3O_4$, $MgFe_2O_4$, and $ZnFe_2O_4$: A combined spectroscopic, diffraction and theory study. *Physical Chemistry Chemical Physics*, 22(45):26200–26215, 2020.
17. Promod Kumar, Mohan Chandra Mathpal, Gajendra Kumar Inwati, Sanjay Kumar, Mart-Mari Duvenhage, Wiets Daniel Roos, and Hendrik C Swart. Study of defect-induced chemical modifications in spinel zinc-ferrites nanostructures by in-depth XPS investigation. *Magnetochemistry*, 9(1):20, 2023.
18. M Sumalatha, S Shravan kumar Reddy, M Sreenath Reddy, Suresh Sripada, M Manivel Raja, Ch Gopal Reddy, P Yadagiri Reddy, and V Raghavendra Reddy. Raman and in-field ^{57}Fe Mössbauer study of cation distribution in Ga substituted cobalt ferrite ($CoFe_2$-$XGaXO_4$). *Journal of Alloys and Compounds*, 837:155478, 2020.
19. Karla R Sanchez-Lievanos, James L Stair, and Kathryn E Knowles. Cation distribution in spinel ferrite nanocrystals: Characterization, impact on their physical properties, and opportunities for synthetic control. *Inorganic Chemistry*, 60(7):4291–4305, 2021.

20. Pallavi G Undre, Ashok V Humbe, Jitendra S Kounsalye, Arun Kumar, RV Kathare, and KM Jadhav. Diamagnetic Al^{3+} doped Ni–Zn spinel ferrite: Rietveld refinement, elastic, magnetic, Mössbauer, and electrical explorations. *Journal of Inorganic and Organometallic Polymers and Materials*, 33(11):3372–3388, 2023.
21. Xin Wang, Brianna L Musicó, Corisa Kons, Peter C Metz, Veerle Keppens, Dustin A Gilbert, Yuanpeng Zhang, and Katharine Page. Local cation order and ferrimagnetism in compositionally complex spinel ferrites. *APL Materials*, 10(12), 2022.
22. Amal Soufi, Hind Hajjaoui, Rachid Elmoubarki, Mohamed Abdennouri, Samir Qourzal, and Noureddine Barka. Spinel ferrites nanoparticles: Synthesis methods and application in heterogeneous Fenton oxidation of organic pollutants – a review. *Applied Surface Science Advances*, 6:100145, 2021.
23. Barbara Pacakova, Simona Kubickova, Alice Reznickova, Daniel Niznansky, and Jana Vejpravova. Spinel ferrite nanoparticles: Correlation of structure and magnetism. *Magnetic Spinels – Synthesis, Properties and Applications*:4–6, 2017.
24. Daliya S Mathew and Ruey-Shin Juang. An overview of the structure and magnetism of spinel ferrite nanoparticles and their synthesis in microemulsions. *Chemical Engineering Journal*, 129(1–3):51–65, 2007.
25. A Broese Van Groenou, PF Bongers, and AL Stuyts. Magnetism, microstructure and crystal chemistry of spinel ferrites. *Materials Science and Engineering*, 3(6):317–392, 1969.
26. AT Ngo, P Bonville, and MP Pileni. Spin canting and size effects in nanoparticles of nonstoichiometric cobalt ferrite. *Journal of Applied Physics*, 89(6):3370–3376, 2001.
27. KL Salcedo Rodríguez, Silvana Jacqueline Stewart, PM Mendoza Zélis, Gustavo Alberto Pasquevich, and CE Rodríguez Torres. Role of defects on the magnetic behaviour of the geometrically frustrated spinel $ZnFe_2O_4$. *Journal of Alloys and Compounds*, 752:289–295, 2018.
28. CE Rodríguez Torres, Gustavo Alberto Pasquevich, P Mendoza Z´elis, Federico Golmar, SP Heluani, Sanjeev K Nayak, Waheed A Adeagbo, Wolfram Hergert, Martin Hoffmann, Arthur Ernst, P Esquinaxi, and SJ Stewart. Oxygen-vacancy-induced local ferromagnetism as a driving mechanism in enhancing the magnetic response of ferrites. *Physical Review B*, 89(10):104411, 2014.
29. Vitaly Zviagin, Chris Sturm, PD Esquinazi, Marius Grundmann, and R¨udiger Schmidt-Grund. Control of magnetic properties in spinel $ZnFe_2O_4$ thin films through intrinsic defect manipulation. *Journal of Applied Physics*, 128(16), 2020.
30. C Upadhyay, HC Verma, V Sathe, and AV Pimpale. Effect of size and synthesis route on the magnetic properties of chemically prepared nanosize $ZnFe_2O_4$. *Journal of Magnetism and Magnetic Materials*, 312(2):271–279, 2007.
31. Changwa Yao, Qiaoshi Zeng, GF Goya, T Torres, Jinfang Liu, Haiping Wu, Mingyuan Ge, Yuewu Zeng, Youwen Wang, and JZ Jiang. $ZnFe_2O_4$ nanocrystals: Synthesis and magnetic properties. *The Journal of Physical Chemistry C*, 111(33):12274–12278, 2007.
32. K Maaz, Arif Mumtaz, SK Hasanain, and Abdullah Ceylan. Synthesis and magnetic properties of cobalt ferrite ($CoFe_2O_4$) nanoparticles prepared by wet chemical route. *Journal of Magnetism and Magnetic Materials*, 308(2):289–295, 2007.
33. Yuqiu Qu, Haibin Yang, Nan Yang, Yuzun Fan, Hongyang Zhu, and Guangtian Zou. The effect of reaction temperature on the particle size, structure and magnetic properties of coprecipitated $CoFe_2O_4$ nanoparticles. *Materials Letters*, 60(29–30):3548–3552, 2006.
34. Zhenfa Zi, Yuping Sun, Xuebin Zhu, Zhaorong Yang, Jianming Dai, and Wenhai Song. Synthesis and magnetic properties of $CoFe_2O_4$ ferrite nanoparticles. *Journal of Magnetism and Magnetic Materials*, 321(9):1251–1255, 2009.

35. M Gich, C Frontera, A Roig, E Taboada, E Molins, HR Rechenberg, JD Ardisson, WAA Macedo, C Ritter, V Hardy, J Sort, V Skumryev, and J Nogu´es. High-and low-temperature crystal and magnetic structures of ε-Fe_2O_3 and their correlation to its magnetic properties. *Chemistry of Materials*, 18(16):3889–3897, 2006.
36. AA Dubrovskiy, DA Balaev, KA Shaykhutdinov, OA Bayukov, ON Pletnev, SS Yakushkin, GA Bukhtiyarova, and ON Martyanov. Size effects in the magnetic properties of ε-Fe_2O_3 nanoparticles. *Journal of Applied Physics*, 118(21), 2015.
37. Mathew George, Asha Mary John, Swapna S Nair, PA Joy, and MR Anantharaman. Finite size effects on the structural and magnetic properties of sol–gel synthesized $NiFe_2O_4$ powders. *Journal of Magnetism and Magnetic Materials*, 302(1):190–195, 2006.
38. Hua Li, Hua-zhong Wu, and Guo-xian Xiao. Effects of synthetic conditions on particle size and magnetic properties of $NiFe_2O_4$. *Powder Technology*, 198(1):157–166, 2010.

2 Synthesis Methods of Nanoferrites

Nancy, Ritesh Verma, Ankush Chauhan, Pradeep Kumar, Preeti Thakur, and Atul Thakur

2.1 INTRODUCTION

The history of ferrite materials starts with the discovery of stones that attract iron. Magnetite (Fe_3O_4) got its name because these stones are so common in the Magnesian area of Asia Minor. One of the magnetite's first uses was in the "lodestones" that early navigators used to find the north of the magnet [1]. In 1600, William Gilbert published the first scientific investigation of magnetism in *De Magnete* [2]. It was in 1819 when Hans Christian Oersted demonstrated the effect of a wire current on a magnetic compass needle [3]. The Latin word "ferrum," which means iron, is where the word "ferrite" originates. Iron oxide is the primary component of ferrites, which are homogeneous ceramic materials made of different oxides [4]. Ferrite materials may be candidates for use in practical applications because they have recently demonstrated modest magnetic characteristics. A variety of metal cations attached to nanostructured metal oxides provide the backbone of the chemical formula for nanoferrite materials, which is AB_2O_4. The amount and position of metal cations in the crystalline structure of ferrites greatly impact their physical and chemical properties. The remarkable properties of nanocrystalline magnetic materials have piqued the curiosity of several scientific fields, including engineering, biology, chemical sciences, physical sciences, and materials science. The large surface-to-volume ratio and particle sizes of nanomaterials (up to 100 nm) set them apart from bulk materials in terms of magnetic, thermal, mechanical, optical, and reactive properties [3 –6]. You could think of nanoferrites as magnetic semiconductors due to their unique combination of ferrimagnetic and semiconductor properties. Among these materials, many electrical and telecommunications-related uses are recording heads, antenna rods, loading coils, microwave devices, and core materials for power transformers [2–4]. Nanosized ferrites have exceptional electric and magnetic properties [5], which are distinct from micro-structured materials and could lead to new uses such as gas sensors, photocatalysis, magnetic drug delivery, high-density data storage, ferrofluids, and more. This study compared the magnetic, thermal, structural, and morphological properties of nanoferrites made by different techniques, including biosynthesis, sol–gel, electrochemical deposition, high-temperature thermal decomposition, microemulsion, and hydrothermal and solvothermal processes [7].

 DOI: 10.1201/9781032665658-2

2.2 SYNTHESIS OF NANOFERRITES

Two processes are required to create nanoparticles with magnetic properties, often used in composite materials. The first step is synthesizing the nanoparticles, and the second step is altering the surface functions of the nanoparticles. Through several methods, it is possible to produce nanoparticles of varying sizes. It is well known that the manufacturing process affects the chemical as well as physical characteristics of nanoparticles after they have been produced. Since the emergence of magnetic nanoparticles, one of the primary areas of study has been conducted is the development of novel preparation techniques and application scenarios. When it comes to the production of magnetic nanoparticles, the approaches that are currently used the most commonly are chemical and physical procedures [8, 9].

The controlled synthesis of various magnetic nanoparticles with varying sizes, contents, and morphologies has been the subject of much research in the literature, which has studied a wide range of synthesis methodologies. This section will focus on the physical, chemical, and biosynthetic procedures that are used the most often in the synthesis of nanoferrites [10–17]. These techniques are the ones that are most employed.

2.2.1 Chemical Methods

As a result of its great capacity to manufacture nanoparticles (NPs) in a fair amount of time and at an affordable cost, chemical synthesis is the method that is most often used for the production of nanoparticles (NPs). To this day, a variety of magnetic nanoparticles have been effectively created by the use of several different approaches that have been presented. All these methods can be divided into two different categories: aqueous solvents and non-aqueous solvents. These two categories can be distinguished from one another in terms of cost, size, and the capacity to regulate the form. Even though they are inexpensive, magnetic nanoparticles that are based on aqueous solutions are challenging to manage in terms of their sizes and forms. The cost of non-aqueous techniques is higher than that of aqueous methods, as they provide more control over the size [18]. This section's primary objective is to provide an overview of the common procedures that are employed in the production of nanoferrite nanoparticles (NPs) and to investigate those procedures.

2.2.1.1 Co-precipitation Method

The co-precipitation method is depicted in Figure 2.1. The method is both quick and inexpensive for the fabrication of spinel ferrite nanoparticles (SFNs). Through this method, nanoferrite nanoparticles are produced that are homogeneous, very pure, and able to be controlled in size [19]. In this method, a base, which acts as a precipitating agent, is combined with a solution that contains stoichiometric concentrations of metallic salts like chlorides, nitrates, or sulfates at temperatures that are suitable for the process. Several parameters have a significant influence on the size, shape, and characteristics of ferrite nanoparticles [20, 21]. These include temperature, pH, the time of the reaction, the kind and ratio of precursors, and other factors.

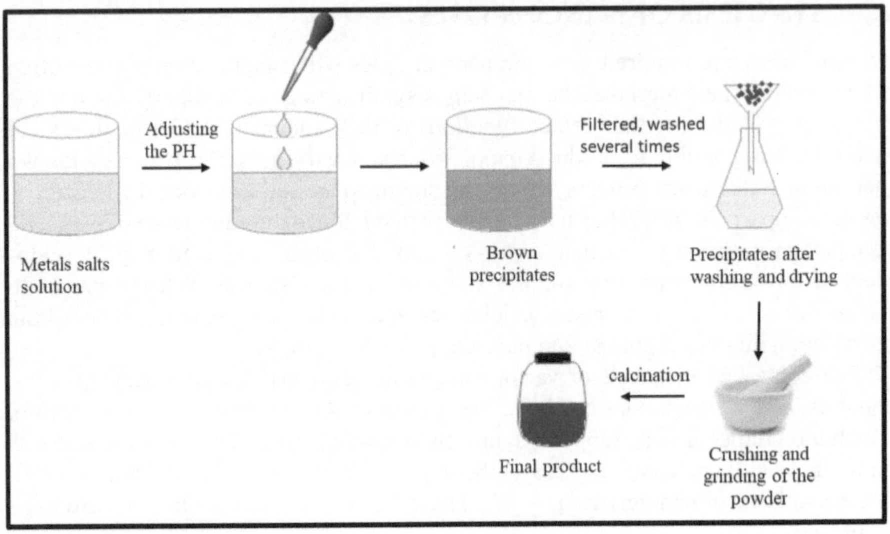

FIGURE 2.1 Synthesis of nanoferrites by co-precipitation method.

This method allows for the production of nanoferrites that have a consistent structure, a high level of purity, and a size that can be adjusted. By using this method, it is possible to effortlessly disseminate a substantial quantity of magnetic nanoparticles (MNPs) in aqueous conditions. The base addition is followed by the precipitation process. This technique is carried out to achieve the desired results. The capacity of the co-precipitation method to manufacture enormous quantities of spinel ferrites and to scale up production is one of the most significant benefits of this technique. The chemical technique known as co-precipitation is the one that is used in the business most often in order to produce hundreds of kilograms of MFe_2O_4 nanoparticles.

However, the major issues with the co-precipitation approach are the size distribution of MNPs, crystalline quality, and morphological control. The key challenge with this technology is the low crystallinity of NPs produced by co-precipitation, which can be addressed with additional heat treatment [22]. In this process, basic components such as nitrate and chloride are dissolved in polar solvents to create a homogeneous solution. The size and structure of nanoparticles can be influenced by conditions including temperature, pH, and the content of precursors that are harmful to crystal development and particle aggregation. Following precipitation, the solid bulk is gathered and washed. After that, calcination is used to create crystalline oxides from the hydroxides.

2.2.1.2 Solid-State Reaction

For the production of nanometric particles, the solid-state reaction technique is frequently used and offers a number of advantages that are not found in other methods. As the solid-state reaction method does not involve the utilization of hazardous solvents or costly organic solvents for the preparation of cobalt ferrite nanoparticles Since the synthesis process takes place at room temperature and at a pressure lower than the atmospheric pressure, it is both economical and straightforward. This method has been used by many researchers in order to produce cobalt ferrite. The fundamental components, namely carbon dioxide (CO_2) and iron oxide (Fe_2O_3), were ground up and used in the right proportions. Additionally, this powdered sample was calcined for 5 h at a temperature of 1100°C in a muffle furnace, with cooling rates of 5°C per minute and heating rates of 10°C per minute. PVA was used as a binder, and the powder was subsequently pelletized into small disks or toroids using the formula [23]. The production of nanoferrites with the required properties can be accomplished by the processes of sintering, calcination, and annealing. The high permeability of cobalt ferrite could be controlled by increasing the amount of energyused in this synthesis. This was achievable because the nanoferrites were generated using an extended sintering technique. One of the many advantages of this method is that it results in lower cost while simultaneously increasing the yield and enhancing the filling qualities. However, it is associated with severe problems, one of which is the generation of hazardous waste products that are not suitable for mass manufacturing.

2.2.1.3 Hydrothermal Method

Crystalline spinel ferrite nanoparticles are produced using hydrothermal synthesis, which involves the use of several wet-chemical processes. MNPs with high purity and variable form may be produced via the use of a hydrothermal technique that is both straightforward and effective, as shown in Figure 2.2. This method of production

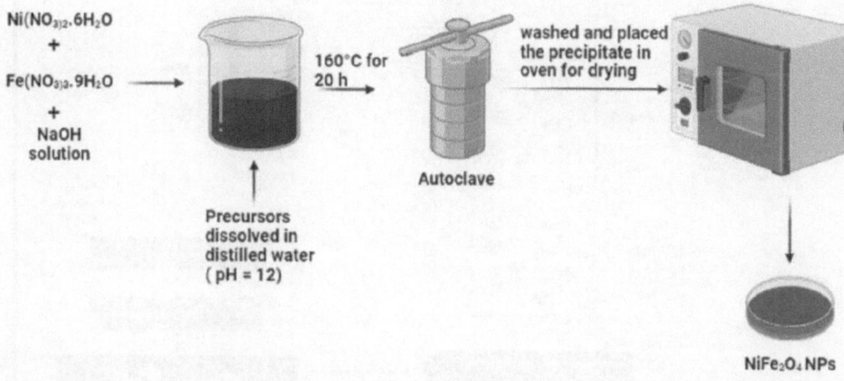

FIGURE 2.2 Schematic diagram of the synthesis of spinel ferrite nanoparticles by hydrothermal route.

involves dissolving and mixing independently soluble salts that comprise divalent and trivalent transition metals at a mole ratio of 1:2, respectively [23]. Continuous stirring is required while organic solvents, such as ethylene glycol or sodium hydroxide, are progressively added to the mixes of aqueous solutions in order to produce a homogeneous solution. An autoclave is then used to bring the solution to a high temperature and pressure. Large-scale production of ferrite nanoparticles may be accomplished by the hydrothermal process, which is one of the most intriguing such methods. In order to produce high-quality nanoferrites with regulated diameters and size distribution, it is possible to vary the reaction duration, temperature, and pressure by selecting the right solvent combination [24–27]. This is a useful point of view since it suggests that changes in reaction temperature, time, dopants, and other aspects can alter particle size, structure, and other physical features [28].

2.2.1.4 Micro-emulsion Method

Three components make up microemulsions: oil, water, and surfactant. These microemulsions are transparent and stable. Surfactants are applied in this method to facilitate the coexistence of two liquids that are incompatible with one another in a single phase. One of the microemulsion solvents used in preparing the solution by dispersing the immiscible solvents is a water/oil-based compound [29]. Direct microemulsion, in which oil is dispersed in water, and reverse microemulsion, in which water is dispersed in oil, are the two microemulsion processes used in the production of magnetic nanoparticles. A graphical representation of the process of producing nanoferrite nanoparticles using the water-in-oil microemulsion method is shown in Figure 2.3. If a monolayer molecule contains a hydrophobic head that is

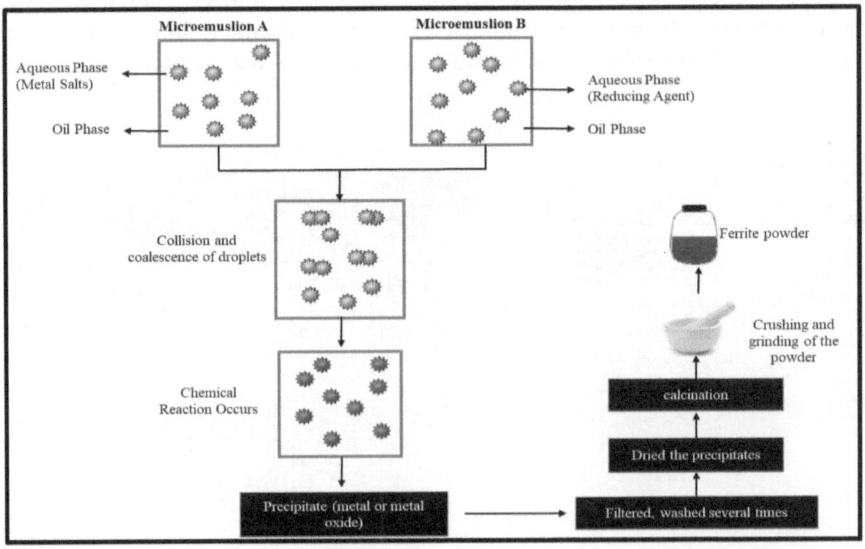

FIGURE 2.3 Ferrites obtained by microemulsion method.

oil soluble and a hydrophilic tail that is water soluble, then it is feasible to employ the monolayer molecule as a surfactant (also known as a surfactant). For instance, dodecyl sulfate is a surfactant used in order to maintain the stability of the solution. The magnetic nanoparticle precursor is commonly dispersed in the aqueous phase as nanodroplets ranging from 1 to 100 nm in size. For the purpose of functioning as nanoreactors, water droplets are encased in "micelles" that are composed of molecules of surfactant.

During the process, magnetic nanoparticles start to grow within the micelles. These nanoparticles enclose the particles and prevent them from multiplying, aggregating, or creating new particles. A second emulsion is added to the solution in order to affect the precipitation of the nanoparticles. Using oil-in-water microemulsion, for instance, to manufacture nickel ferrite (NPs) as photocatalysts for the separation of water allowed for the development of nanoferrite NPs with diameters that were less than 20 nm. This method has the advantage of being able to readily size-modulate MFe_2O_4 MNPs to sub-nanometer sizes, which makes it possible to quickly modify the sizes of the particles. The microemulsion method, on the other hand, allows for the production of a large variety of distinct shapes and size distributions. This method, in contrast to others, has a limited window of operation and a poor yield. It is also more difficult to implement. On the other hand, a number of studies [8, 18, 30] have yielded favorable findings that support the idea of expanding the process for a facility that focuses on either the economics or the environment.

2.2.1.5 Thermal Decomposition Method

The size and morphological limitations of the co-precipitation technique are circumvented by the thermal degradation technique. Magnet nanoparticles with more consistent sizes are often the result of heating the material to a higher temperature. Another option for producing more crystalline MFe_0O_4 MNPs is high-temperature breakdown. Magnetic nanoparticles with a highly crystalline structure, homogeneous shape, and limited size distribution may be synthesized by separating the nucleation of nanoparticles and growth phases rather than co-precipitation [31]. Size selection strategies are unnecessary for the creation of monodisperse magnetic nanoparticles. The method may generate nanoferrite NPs of many forms, including cubic and octahedral, by adjusting the heating rate, temperature, concentration of organic compounds, and type of organic solvent [32, 33]. Due to the narrow size range and high crystallinity of nanoferrite NPs, thermal breakdown is better than co-precipitation production for biomedical uses [34].

2.2.1.6 Sol–Gel Method

Chemically, the sol–gel technique [35] may be used to synthesize single and multicomponent materials into exceedingly tiny particles at low temperatures. Chemical and physical processes including polymerization, densification, gelation, condensation, drying, and hydrolysis are all parts of the "sol–gel" multi-step process. The sol–gel synthesis of ferrite is shown in Figure 2.4. The sol–gel method involves the gelation of materials to be recovered from chemical solutions. The first is the inorganic approach, which employs metal salts in water (chloride, nitrate), and the second

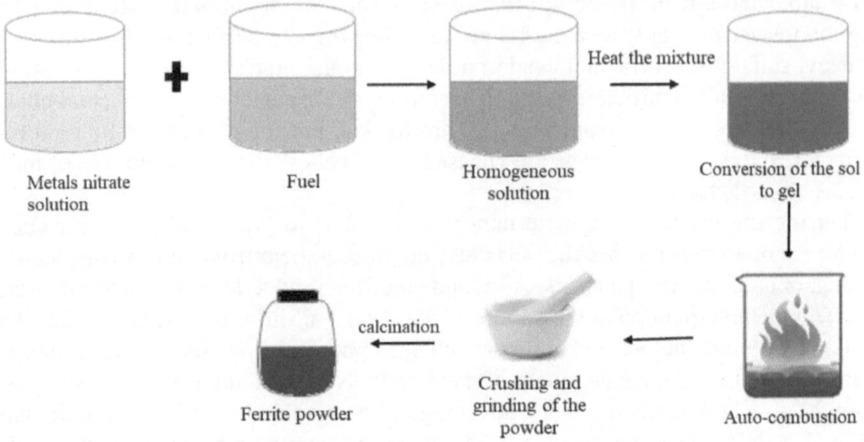

FIGURE 2.4 Synthesis of nanoferrites by sol–gel auto combustion method.

is the metal-organic way, which employs metal alkoxides in oil. It is more difficult to regulate the reactions using the inorganic technique, but it is cheaper and easier to handle than using metal alkoxides. This method allows for more controllable phase development, consistent particle size, and the desired stoichiometry [36–38]. Since sol–gel processing creates a homogeneous distribution of materials within the gel, there is no potential of contamination, in contrast to the solid-state reaction approach, which involves grinding processes to generate homogeneous single-phase ceramics. Some disadvantages of the sol–gel approach include the following: a long processing time, difficulties with phase separation, high carbon content in the final products, several processes, careful process monitoring required, and costly raw ingredients compared to metal ion sources derived from minerals [39].

2.2.1.7 Electrochemical Method

To put it simply, the essential concept behind an electrochemical synthesis is the exchange of an electric current between two or more electrodes (anode and cathode) that are positioned in an electrolyte. The metal ion species that the anode has produced in the electrolyte may be converted into a metal by the cathode, provided that the cathode is supported by stabilizers [40]. M. Rivero and colleagues [41] conducted research on the primary method by which $Zn_xFe_{3-x}O_4$ ferrite nanoparticles may be produced through electrochemical synthesis. Electrochemical methods were used by *Ovejero* et al. [42] in order to produce manganese, cobalt, and nickel ferrites. Based on the findings of the study, it has been shown that the ability of magnetic nanoparticles to form films is significantly influenced by the composition of the electrolyte as well as the deposition potential.

2.2.1.8 Sonochemical Method

Several sonochemical methods make use of the chemical reactions generated by sonic cavitation to create metal-ferrite magnetic nanoparticles with varying shapes

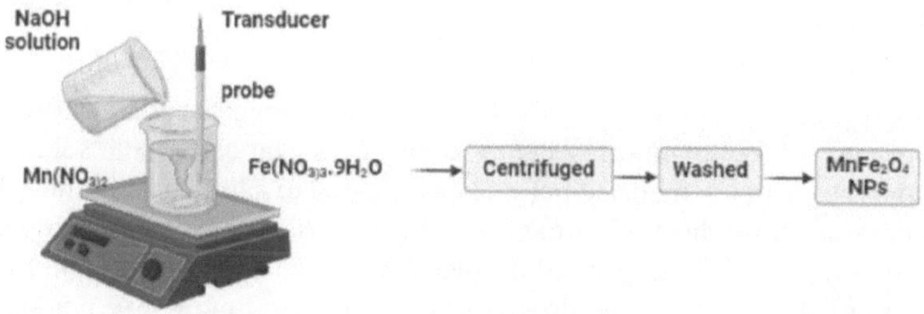

FIGURE 2.5 Synthesis of nanoferrites by sonochemical method.

and sizes [43]. The ultrasonic irradiation causes the bubbles to constantly compress and expand, causing them to oscillate. A progressive expansion causes the pulsing bubbles that hold the ultrasonic energy to rupture, releasing it. The rapid and dramatic increase in temperature and pressure is caused by the highly concentrated energy released when the bubbles burst [44]. The production of magnetic nanoparticles with a limited size range is challenging to the fascinating prospect of sonolysis or sonochemical creation of these particles, which might lead to their extraordinary magnetic saturation capabilities. This process results in an abundance of magnetic nanoparticles that are aggregated, porous, and amorphous [18, 45]. The average size of the crystallites in $MnFe_2O_4$ NPs grew from 1.8 to 22.1 nm when the sonication duration was extended from 20 to 80 min, as reported by Yadav et al. [45]. Using an 80-min sonication duration, Figure 2.5 shows the sonochemical synthesis of $MnFe_2O_4$ NPs. Cylindrical-shaped nanostructures of $Zn_{0.35}Fe_{2.65}O_4$ with an average size of 10–40 nm were successfully synthesized using the sonochemistry method and PEG-6000 as a surfactant, and then calcined at 500°C for 2 h [46]. Another work used ultrasonic irradiation to produce Mn-Zn nanoferrites doped with Er^{3+} and Dy^{3+} [47].

2.2.1.9 Ceramic Method

The classic ceramic technique may be used to produce a wide variety of compounds, including nitride, aluminosilicate, sulfide, and mixed metal oxides. In order to produce a mixture that is homogeneous, it is necessary to first combine two or more solid compounds, which is referred to as homogenization. Subsequently, the mixture must be ground in a wet medium in order to regulate the particle size, then it must be compressed and heated to a high temperature [48, 49]. However, in order to bypass the lattice energy barrier and enable cations and anions to diffuse into a new place, this method requires a high temperature (between 500 and 2000°C). This is because it requires a significant amount of energy to do this. The anticipated chemical will decompose when subjected to high temperatures. It is generally true that the pace of the reaction is slow; however, as the temperature increases, the rate of the reaction quickens because the diffusion rate also increases, which prevents the solid form from melting. In a significant way, the features of SFs and HFs are influenced by the synthesis process that was utilized. Without contaminants and with little particle aggregation, it is challenging to produce a powder that is fine, uniform, and free of impurities.

2.2.1.10 Pyrolysis

2.2.1.10.1 Spray Pyrolysis Method

During the process of spray pyrolysis, the mixture of reagents is transformed into droplets of aerosol. Following this step, the solvent is evaporated, the solute is condensed, and the particles are dried before being heated to a high temperature. For the purpose of controlling the environment in which particles are formed, this method involves dividing the fluid into small droplets. As a result of ensuring the full preservation of stoichiometry on the droplet scale and providing a very homogeneous distribution of the components, it is often suitable for the synthesis of mixed metal ferrites [50]. Through the regulation of the type of thermolysis reaction, the type of precursors, the gaseous carrier, the deposition length, and the substrate temperature, this technology makes it feasible to produce a broad range of hollow or porous particles that have the potential to be used in thermal insulation or catalyst support [50, 51].

Within the reaction mixture, the addition of soluble neutral additives (NaCl, H_3BO_3) may result in the formation of non-agglomerated particles that are smaller than 10 nm. In addition, the inert components make it possible to create further heat processing of the powder without causing any significant increase in particle size [52]. Some of the benefits of the technology include a high production rate, the capacity to scale, and processing times that are significantly reduced [44, 52]. Additionally, it generates particles of consistent size and fine dispersion, which might vary in both shape and composition. One of the method's downsides is that it might potentially result in the accumulation of particles, in addition to having high expenses. The process of spray pyrolysis, which involves spraying an aqueous metal nitrate solution at a temperature of 300°C, was used to form thick layers of $Zn_xFe_{3-x}O_4$ and $CoFe_2O_4$ [53]. The film was spontaneously cooled after it was first deposited. The spray pyrolysis approach is used for the preparation of $CoFe_2O_4$ thin films [53]. This method is utilized because it is an easy experimental setup, is cost-effective, and has the capability to build extensive surfaces of thin film.

2.2.1.10.2 Laser Pyrolysis

Another technique for creating ferrites in the nanometer range is laser pyrolysis. A carrier gas is used to carry the gaseous phase precursors into a reaction chamber, which produces raised local temperatures as a result of a high-power laser beam (2400 W), causing NPs to form and develop before being further collected by a filter [54]. Vapor or tiny droplets of liquid reactants are used to introduce them into the reaction chamber [54]. Using this technology, high-purity nanoparticles with a small and narrow range of particle sizes can be made in one step. The process is flexible and adaptive, enabling the adjustment of operating parameters such as gas type, laser power, gas flow rate, and the concentration of precursor [55].

Additionally, it is simple to scale up to industrial manufacturing. A precursor exhibiting an absorption band at 10.6 m is required, and adding C_2H_4 or NH_3 increases the amount of nitrogen or carbon contamination [54, 55]. These are some of the disadvantages of the process. It was discovered that $ZnFe_2O_4$ nanopowders may be created by laser pyrolysis using zinc and iron nitrate solutions as starting reagents, ethylene as the sensitizer gas, and air as the carrier gas. The resulting nanopowders were put to the test as Li-ion battery-negative electrode materials.

2.2.1.11 Polyol Method

By using a solvent with a relatively high boiling point that also decreases metallic ions in reflux situations, the polyol method reduces metal oxides or metallic complexes. Precursors are dissolved in liquid polyol (ethylene, diethylene, and triethylene glycol), heated to the boiling point of the polyol, and then cooled to regulate particle growth and avoid particle aggregation. The key benefits of this process are the uniformly sized soft and hard magnetic NPs that are produced, whereas the main drawbacks are the high temperature and extended duration [56, 57]. Numerous nanoparticles have been synthesized using the polyol approach [58]. To synthesize $CoFe_2O_4$, $FeCl_3{\cdot}6H_2O$, and $CoCl_2{\cdot}4H_2O$ were both independently dissolved in glycol and then combined while being constantly stirred. Water, sodium acetate, and ammonia solution were all added to the homogeneous solution, and the pH of the mixture was then changed to 8, 10, and 12 by the addition of the NH_3 solution. The mixture was brought to the boiling point of glycol under reflux conditions, and the NPs that were created were separated by centrifugation. Moreover, formamide, citric acid, and PVA were employed as modifiers in the $NiFe_2O_4$ nanocomposite synthesis, which began with tetraethoxysilane [59].

2.2.1.12 Biosynthesis

The manufacture of nanoferrites using a microbial enzyme or a phytochemical from plants that has the ability to reduce and/or stabilize is a new topic of study that presents a viable alternative to chemical and physical methods [60, 61]. Nanoferrites have traditionally been produced using magnetotactic and iron-reducing bacteria. By doping metal ferrite nanoparticles with cobalt, this technique has the potential to enhance their magnetic characteristics. The type of bacteria used throughout the preparation process (aerobic or anaerobic) and their phase determine the final magnetic nanoparticles that are created. Actinobacteria, for example, can be used to make superparamagnetic maghemite nanoparticles in an aerobic atmosphere [62]. It is challenging to identify the parameters that control size and shape while maintaining the appropriate saturation magnetization because the biogenesis mechanism of magnetic nanoparticles is frequently poorly known.

2.2.2 Physical Method

In order to disintegrate bulk ferrite materials into nanoparticles, the majority of the physical procedures that are used are top-down approaches. Explosions of electrical wires, mechanical grinding, and the evaporation of laser targets are a few examples of well-known physical techniques [63–65]. This is due to the fact that physical operations tend to use a great deal of power and are challenging to regulate in terms of the distribution of form and size. The generation of nanoferrite nanostructures was first accomplished by the use of mechanical ball milling [66]. In this process, high-energy ball mills are used to grind powders (such as oxides, carbonates, and other metal-containing compounds).. The resultant mixture is then subjected to high temperatures in order to transform it into the required material phase. Furthermore, the mechanical ball milling process may be carried out in two distinct variations: dry and wet. The wet method is an excellent technique for producing crystal particles that have a considerable amount of magnetism. Despite of the fact that the ball milling technique is straightforward to implement, the crystal structure of MFe_2O_4 may be influenced by both mechanical and chemical factors, and it is easy to include impurities that are present during the preparation stage. Because of this, the ball milling method is not effective in producing nanoferrite crystals with a variety of morphologies.

Nanoscale ferrites have gained a lot of attention in recent years due to their unique properties. These properties include the ability to maintain their chemical and thermal stability as well as significant coercivity, elevated anisotropy constant, and Curie temperature. Further, moderate saturation magnetization, high electrical resistance, and minimal loss of eddy current are also maintained. The beneficial features of magnetic nanoferrites, when combined with typical modification and functionalization, bring up new options for their future use in the fields of biomedicine, catalysis, wastewater treatment, and energy storage. One of the most common applications of MFe_2O_4 is seen in Figure 2.6. The features of ferrite may alter depending on the kind

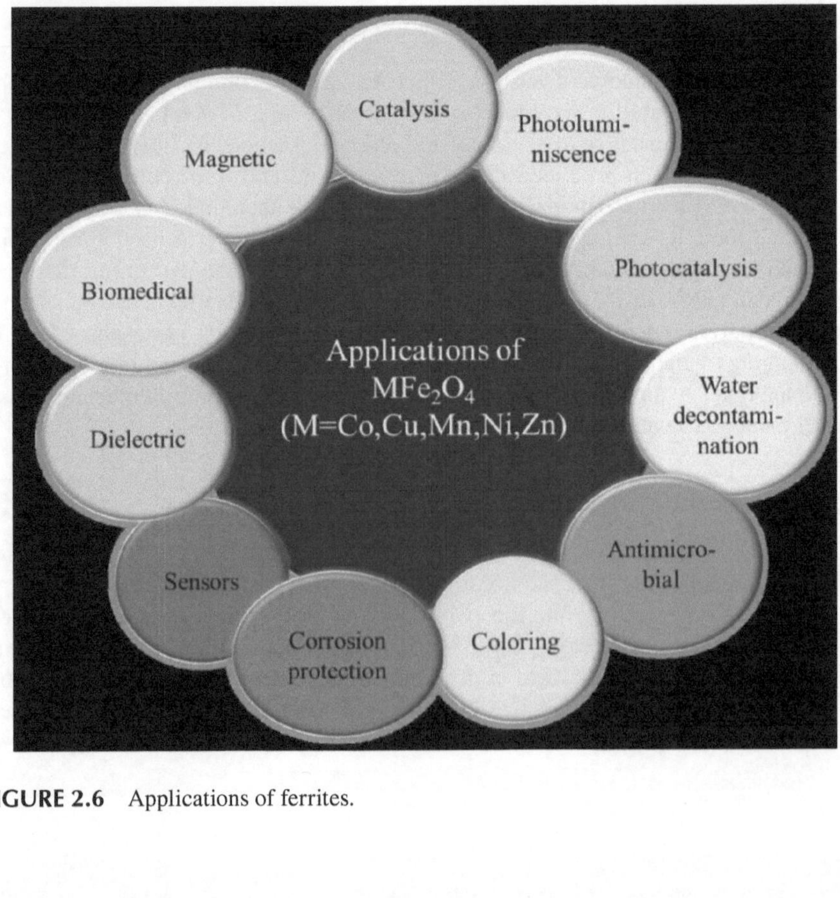

FIGURE 2.6 Applications of ferrites.

of cation and the synthesis procedure utilized. This can result in a wide range of uses for ferrite. In addition to synthesis techniques and modification, doping nanoferrites is an additional effective strategy for influencing the physical and chemical characteristics of nanoferrites. This method allows for the formation of nanoferrites that are monodisperse and non-aggregated. As an effective magnetic material, magnetic nanoferrites have the potential to be used in the treatment of magnetic hyperthermia as well as in a wide variety of other biomedical sectors. Although a significant amount of research has been conducted on the use of nanomaterials based on ferrite for the treatment of hyperthermia, the field is still in its infancy. Additional work is required to address specific problems, such as the modification of size, structure, and magnetic characteristics of nanoparticles.

To generate magnetic ferrite nanoparticles with the performance that is predicted for use in biomedical applications, one of the most important challenges that must be overcome is developingeffective methods for modifying the surface of nanoferrite components. Even if the number of applications of magnetic nanoferrites for wastewater treatment has increased, there is still a need for more research to address several challenges to improve the photocatalytic capabilities of ferrite nanoparticles for the destruction of organic pollutants.

2.3 CONCLUSION

The production of nanoferrites, which belong to a category of magnetic materials that include particles as tiny as nanometers, has garnered a significant amount of interest in the field of study due to their unique magnetic properties and numerous applications that it has in various sectors, including environmental cleanup, biomedicine, and electronics development. The size, shape, and content of nanoferrites may be controlled using a range of synthesis processes that have been developed and optimized. These techniques include sol–gel, hydrothermal, co-precipitation, and microwave-assisted synthesis, among others. Regarding the homogeneity of particles, crystallinity, and scalability, each technique has its own set of distinct benefits. The magnetic properties of nanoferrites, which include low coercivity, high saturation magnetization, and superparamagnetism, are significantly impacted by their size, shape, and the specific synthesis procedure that is used to create them. By making adjustments to these qualities, nanoferrites may be adapted to meet the requirements of certain applications. Despite the significant progress that has been made, there are still challenges that need to be conquered to manufacture nanoferrites on a wide scale, keep their quality consistent, and understand their long-term stability and toxicity. The objectives of the research that will be conducted in the future are to find solutions to these problems, examine novel methods of synthesis, and broaden the spectrum of uses for nanoferrites.

REFERENCES

1. A.A. Mills, The lodestone: History, physics, and formation, *Ann. Sci.* 61 (2004) 273e319.

2. S. Malin, D. Barraclough, G. De Magnete, An early study of magnetism and electricity, *Eos Trans. Am. Geophys. Union.* 81 (2000) 233e234.
3. S. Stringari, R.R. Wilson, Romagnosi and the discovery of electromagnetism, *Rend. Lincei.* 11 (2000) 115e136.
4. M. Kaur, N. Kaur, Vibha, Ferrites: Synthesis and applications for environmental remediation, *Ferrites Ferrates Chem. Appl. Sustain. Energy Environ. Remediat.* (2016) 113–136.
5. P. Kumar, S.K. Sharma, M. Knobel, M. Singh, Effect of La3þ doping on the electric, dielectric and magnetic properties of cobalt ferrite processed by co-precipitation technique, *J. Alloys Compd.* 508 (2010) 115–118.
6. F. Gomes da Silva, J. Depeyrot, A. Campos, R. Aquino, D. Fiorani, D. Peddis, Structural and magnetic properties of nano ferrite nanoparticles, *J. Nanosci. Nanotechno.* 19 (2019) 1–15. https://doi.org/10.1166/jnn.2019.16877.
7. S. Shaikh, M. Ubaidullah, R. Mane, A. Al-Enizi, Types, synthesis methods and applications of ferrites (2020). https://doi.org/10.1016/B978-0-12-819237-5.000043.
8. S. Liu, B. Yu, S. Wang, Y. Shen, H. Cong, Preparation, surface functionalization and application of Fe_3O_4 magnetic nanoparticles, *Adv. Colloid Interface Sci.* 281 (2020) 102165.
9. A. Allafchian, S.S. Hosseini, Antibacterial magnetic nanoparticles for therapeutics: A review, *IET Nanobiotechnol.* 13 (8) (2019) 786–799.
10. M. Vishwas, K.V. Babu, K.A. Gowda, S.B. Gandla, Synthesis, characterization and photo-catalytic activity of magnetic $CoFe_2O_4$ nanoparticles prepared by temperature controlled co-precipitation method, *Mater. Today: Proc.* 9 (6) (2022) 16601.
11. P. Thakur, N. Gahlawat, P. Punia, S. Kharbanda, B. Ravelo, A. Thakur. Cobalt nanoferrites: A review on synthesis, characterization, and applications, *Journal of Superconductivity and Novel Magnetism.* 35 (10) (2022) 2639–2669.
12. A.A. Rodríguez-Rodríguez, M.B. Moreno-Trejo, M.J. Melendez-Zaragoza, V. Collins-Martínez, A. Lopez-Ortiz, E. Martínez-Guerra, M. Sanchez-Domínguez, Nano-type ferrite nanoparticles: Synthesis by the oil-in-water microemulsion reaction method and photocatalytic water-splitting evaluation, *Int. J. Hydrogen Energy* 44 (24) (2019) 12421–12429.
13. G. Allaedini, S.M. Tasirin, P. Aminayi, Magnetic properties of cobalt ferrite synthesized by hydrothermal method. *Int Nano Lett* 5 (2015) 183–186. https://doi.org/10.1007/s40089-015-0153-8
14. Z. Mahhouti, H. Moussaoui, T. Mahfoud, M. Hamedoun, M.E.L. Marssi, A. Lahmar, A. Kenz, A. Benyoussef, Chemical synthesis and magnetic properties of monodisperse cobalt ferrite nanoparticles, *Journal of Materials Science: Materials in Electronics.* 30 (2019) 4913–14922.
15. S. Kanagesan, M. Hashim, S. Ab Aziz, I. Ismail, S. Tamilselvan, N.B. Alitheen, M.K. Swamy, B. Purna Chandra Rao, Evaluation of antioxidant and cytotoxicity activities of copper ferrite ($CuFe_2O_4$) and zinc ferrite ($ZnFe_2O_4$) nanoparticles synthesized by sol-gel self-combustion method, *Appl. Sci.* 6 (9) (2016) 184.
16. R.S. Yadav, I. Kuritka, J. Vilcakova, J. Havlica, L. Kalina, P. Urbanek, M. Machovsky, D. Skoda, M. Masar, M. Holek, Sonochemical synthesis of Gd^{3+} doped $CoFe_2O_4$ Nano ferrite nanoparticles and its physical properties, *Ultrason. Sonochem.* 40 (2018) 773–783.
17. Salih SJ, Mahmood WM. Review on magnetic Nano ferrite (MFe_2O_4) nanoparticles: From synthesis to application. *Heliyon.* 9(6) May 26 (2023) e16601.

18. Z. Mahhouti, H. El Moussaoui, T. Mahfoud, M. Hamedoun, M. El Marssi, A. Lahmar, A. El Kenz, A. Benyoussef, Chemical synthesis and magnetic properties of monodisperse cobalt ferrite nanoparticles, *J. Mater. Sci. Mater. Electron.* 30 (16) (2019) 14913–14922.
19. P. Thakur, D. Chahar, S. Taneja, N. Bhalla, A. Thakur, A review on MnZn ferrites: Synthesis, characterization and applications, *Ceram. Int.* 46 (10) (2020) 15740–15763.
20. C. Sedrati, S. Alleg, H. Boussafel, A. Bendali Hacine, Structure and magnetic properties of nickel ferrites synthesized by a facile co-precipitation method: Effect of the Fe/Ni ratio, *J. Mater. Sci. Mater. Electron.* 32 (19) (2021) 24548–24559.
21. Y. Peng, C. Xia, M. Cui, Z. Yao, X. Yi, Effect of reaction condition on microstructure and properties of (NiCuZn) Fe2O4 nanoparticles synthesized via co-precipitation with ultrasonic irradiation, *Ultrason. Sonochem.* 71 (2021), 105369.
22. C.R. Kalaiselvan, S.S. Laha, S.B. Somvanshi, T.A. Tabish, N.D. Thorat, N.K. Sahu, Manganese ferrite ($MnFe_2O_4$) nanostructures for cancer theranostics, *Coord. Chem. Rev.* 473 (2022) 214809.
23. K.C. Babu Naidu, W. Madhuri, Hydrothermal synthesis of $NiFe_2O_4$ nano-particles: Structural, morphological, optical, electrical and magnetic properties. Bull Mater Sci. 40(2) (2017) 417–425. https://doi.org/10.1007/s12034-017-1374-4
24. M. Sanna Angotzi, V. Mameli, C. Cara, V. Grillo, S. Enzo, A. Musinu, C. Cannas, Defect-assisted synthesis of magneto-plasmonic silver-nano ferrite heterostructures in a flower-like architecture, *Sci. Rep.* 10 (1) (2020) 1–13.
25. Z. Rouhani, J. Karimi-Sabet, M. Mehdipourghazi, A. Hadi, A. Dastbaz, Response surface optimization of hydrothermal synthesis of Bismuth ferrite nanoparticles under supercritical water conditions: Application for photocatalytic degradation of Tetracycline, *Environ. Nanotechnol. Monit. Manag.* 11 (2019) 100198.
26. S. Fayazzadeh, M. Khodaei, M. Arani, S. Mahdavi, T. Nizamov, A. Majouga, Magnetic properties and magnetic hyperthermia of cobalt ferrite nanoparticles synthesized by hydrothermal method, *J. Supercond. Nov. Magnetism.* 33 (7) (2020) 2227–2233.
27. K. Rajashekhar, J.L. Naik, M. Prasad, C. Venkateshwarlu, K.M. Kumar, Brief review on the effect of synthesis method and doping on the structure and magnetic properties of zinc ferrites, *International Journal of Engineering Research and Applications.* 10 (7) (2020) 24–33.
28. F.D.A. La Porta, C.A. Taft, Emerging research in science and engineering based on advanced experimental and computational strategies (2020).
29. M. Salvador, G. Gutiérrez, S. Noriega, A. Moyano, M.C. Blanco-López, M. Matos, Microemulsion synthesis of superparamagnetic nanoparticles for bioapplications. *Int. J. Mol. Sci.* 22(1) Jan 4 (2021):427.
30. F. Sharifianjazi, M. Moradi, N. Parvin, A. Nemati, A.J. Rad, N. Sheysi, A. Abouchenari, A. Mohammadi, S. Karbasi, Z. Ahmadi, Magnetic $CoFe_2O_4$ nanoparticles doped with metal ions: A review, *Ceram. Int.* 46 (11) (2020) 18391–18412.
31. S. Gul, S.B. Khan, I.U. Rehman, M.A. Khan, M. Khan, A comprehensive review of magnetic nanomaterials modern day theranostics, *Front. Mater.* 6 (2019) 179.
32. R. Chandunika, R. Vijayaraghavan, N.K. Sahu, Magnetic hyperthermia application of $MnFe_2O_4$ nanostructures processed through solvents with the varying boiling point, *Mater. Res. Express.* 7 (6) (2020) 064002.
33. P.B. Balakrishnan, N. Silvestri, T. Fernandez-Cabada, F. Marinaro, S. Fernandes, S. Fiorito, M. Miscuglio, D. Serantes, S. Ruta, K. Livesey, Exploiting unique alignment of cobalt ferrite nanoparticles, mild hyperthermia, and controlled intrinsic cobalt toxicity for cancer therapy, *Adv. Mater.* 32 (45) (2020) 2003712.

34. K.K. Kefeni, T.A. Msagati, T.T. Nkambule, B.B. Mamba, Nano ferrite nanoparticles and nanocomposites for biomedical applications and their toxicity, *Mater. Sci. Eng. C* 107 (2020) 110314.
35. E. Kianfar, W. Suksatan, Nanomaterial by sol-gel method: Synthesis and application. advances in materials science and engineering, *Advances in Materials Science and Engineering*. 1 (2021) 5102014.
36. A. Šutka, G. Mezinskis, Sol–gel auto-combustion synthesis of nano-type ferrite nanomaterials, *Front. Mater. Sci.* 6 (2012) 128–141.
37. A. Varma, A.S. Mukasyan, A.S. Rogachev, K.V. Manukyan, Solution combustion synthesis of nanoscale materials, *Chem. Rev.* 116 (23) (2016) 14493–14586.
38. A.S. Mukasyan, P. Epstein, P. Dinka, Solution combustion synthesis of nanomaterials, *Proc. Combust. Inst.* 31 (2) (2007) 1789–1795.
39. R.R. Bhosale, A. Kumar, F. AlMomani, I. Alxneit, Propylene oxide assisted sol–gel synthesis of zinc ferrite nanoparticles for solar fuel production, *Ceram. Int.* 42 (2) (2016) 2431–2438.
40. D. Ramimoghadam, S. Bagheri, S.B. Abd Hamid, Progress in electrochemical synthesis of magnetic iron oxide nanoparticles, *J. Magn. Magn Mater.* 368 (2014) 207–229.
41. M. Rivero, A. del Campo, A. Mayoral, E. Mazario, J. Sanchez-Marcos, A. Munoz-Bonilla, Synthesis and structural characterization of $Zn_xFe_{3-x}O_4$ ferrite nanoparticles obtained by an electrochemical method, *RSC Adv.* 6 (46) (2016) 40067–40076.
42. J.G. Ovejero, A. Mayoral, M. Canete, M. García, A. Hernando, P. Herrasti, Electrochemical synthesis and magnetic properties of MFe_2O_4 (M = Fe, Mn, Co, Ni) nanoparticles for potential biomedical applications, *J. Nanosci. Nanotechnol.* 19 (4) (2019) 2008–2015.
43. A.M. Ilosvai, D. Dojcsak, C. Varadi, M. Nagy, F. Kristaly, B. Fiser, B. Viskolcz, L. Vanyorek, Sonochemical combined synthesis of nickel ferrite and cobalt ferrite magnetic nanoparticles and their application in glycan analysis, *Int. J. Mol. Sci.* 23 (9) (2022) 5081.
44. J.H. Bang, K.S. Suslick, Applications of ultrasound to the synthesis of nanostructured materials, *Adv. Mater.* 22 (10) (2010) 1039–1059.
45. R.S. Yadav, I. Kuritka, J. Vilcakova, T. Jamatia, M. Machovsky, D. Skoda, P. Urbanek, M. Masar, M. Urbanek, L. Kalina, Impact of sonochemical synthesis condition on the structural and physical properties of $MnFe_2O_4$ nano ferrite nanoparticles, *Ultrason. Sonochem.* 61 (2020) 104839.
46. S.R. Yousefi, O. Amiri, M. Salavati-Niasari, Control sonochemical parameter to prepare pure $Zn0.35Fe_2.65O_4$ nanostructures and study their photocatalytic activity, *Ultrason. Sonochem.* 58 (2019) 104619.
47. M. Sertkol, Y. Slimani, M. Almessiere, H. Sozeri, R. Jermy, A. Manikandan, S. Shirsath, A. Ui-Hamid, A. Baykal, Sonochemical synthesis of Mn0.5Zn0. $5Er_xDy_xFe_{2-2x}O_4$ ($x \leq 0.1$) nano nano ferrites: Magnetic and textural investigation, *J. Mol. Struct.* 1258 (2022) 132680.
48. C.N.R. Rao, K. Biswas, Ceramic methods, *Essentials of Inorganic Materials Synthesis*, John Wiley & Sons, 2015, 17–21.
49. A. Vedrtnam, K. Kalauni, S. Dubey, A. Kumar, A comprehensive study on structure, properties, synthesis and characterization of ferrites, *AIMS Mater. Sci.* 7(6) (2020) 800–835. https://doi.org/10.3934/matersci.2020.6.800
50. N. Ajinkya, X. Yu, P. Kaithal, H. Luo, P. Somani, S. Ramakrishna, Magnetic iron oxide nanoparticle (IONP) synthesis to applications: Present and future. *Materials.* 13(20) (2020) 4644.

51. D. Kotsikau, M. Ivanovskaya, V. Pankov, Y. Fedotova, Structure and magnetic properties of manganese zinc-ferrites prepared by spray pyrolysis method. *Solid State Sci.* 39 (2015) 69–73.
52. D. Peddis, C. Cannas, A. Musinu, A. Ardu, F. Orru, D. Fiorani, S. Laureti, D. Rinaldi, G. Muscas, G. Concas, et al., Beyond the effect of particle size: Influence of $CoFe_2O_4$ nanoparticle arrangements on magnetic properties. *Chem Mater.* 25 (2013) 2005–2013.
53. A.V. Jundale, G.Y. Chorage, A.A. Yadav, Structural, morphological and optical properties of $CoFe_2O_4$ thin film by spray pyrolysis technique. *Mater. Today Proc.* 43 (2021) 2678–2681.
54. S. Wang, L. Gao, Laser-Driven nanomaterials and laser-enabled nanofabrication for industrial applications, in: *Industrial Applications of Nanomaterials*, T. Sabu, G. Yves, B.P. Yasir, Eds. Elsevier: Amsterdam, The Netherlands, 2019, pp. 181–203.
55. S. Bourrioux, L.P. Wang, Y. Rousseau, P. Simon, A. Habert, Y. Leconte, T.M. Sougrati, L. Stievano, L. Monconduit, Z.J. Xu, et al., Evaluation of electrochemical performances of $ZnFe_2O_4/\gamma\text{-}Fe_2O_3$ nanoparticles prepared by laser pyrolysis. *New J. Chem. R. Soc. Chem.* 41 (2017) 9236–9243.
56. A.G. Nene, M. Takahashi, K. Wakita, M. Umeno, Size controlled synthesis of Fe3O4 nanoparticles by ascorbic acid mediated reduction of $Fe(acac)_3$ without using capping agent, *J. Nano Res.* 40 (2016) 8–19.
57. S.F. Hasany, N.H. Abdurahman, A.R. Sunarti, R. Jose, Magnetic iron oxide nanoparticles: Synthesis and applications. *Curr. Nanosci.* 9 (2013) 1–15.
58. M. Eslamian, M. Ahmed, N. Ashgriz, Modelling of nanoparticle formation during spray pyrolysis. *Nanotechnology* 17 (2006) 1674–1685.
59. D.S. Jung, Y.N. Ko, Y.C. Kang, S. Bin Park, Recent progress in electrode materials produced by spray pyrolysis for next-generation lithium ion batteries. *Adv. Powder Technol.* 25 (2014) 18–31.
60. S. Kalia, A. Kumar, N. Munjal, N. Prasad, Synthesis of ferrites using various parts of plants: A mini review, in: *Journal of Physics: Conference Series*, 1964 (3) (2021) 032003.
61. P. Thakur, S. Taneja, D. Chahar, B. Ravelo, A. Thakur, Recent advances on synthesis, characterization and high frequency applications of Ni-Zn ferrite nanoparticles, *J. Magn. Magn Mater.* 530 (2021) 167925.
62. A.A. Bharde, R.Y. Parikh, M. Baidakova, S. Jouen, B. Hannoyer, T. Enoki, B. Prasad, Y.S. Shouche, S. Ogale, M. Sastry, Bacteria-mediated precursor-dependent biosynthesis of superparamagnetic iron oxide and iron sulfide nanoparticles, *Langmuir* 24 (11) (2008) 5787–5794.
63. A. Fomenko, A. Kondranova, S. Kazantsev, A. Lozkomoev, A. Pervikov, Antimicrobial activity of CuFe2O4 nanoparticles obtained by electric explosion of Fe and Cu wires, in: *AIP Conference Proceedings*, 2167 (1) 2019. AIP Publishing LLC.
64. E. Glazkova, O. Bakina, N. Rodkevich, A. Mosunov, E. Vornakova, V. Chzhou, M. Lerner, Copper ferrite/copper oxides (I, II) nanoparticles synthesized by electric explosion of wires for high performance photocatalytic and antibacterial applications, *Mater. Sci. Eng., B* 283 (2022) 115845.
65. J.L. Guo, Y.D. Chiou, W.I. Liang, H.J. Liu, Y.J. Chen, W.C. Kuo, C.Y. Tsai, K.A. Tsai, H.H. Kuo, W.F. Hsieh, Complex oxide–noble metal conjugated nanoparticles, *Adv. Mater.* 25 (14) (2013) 2040–2044.
66. S. Liu, B. Yu, S. Wang, Y. Shen, H. Cong, Preparation, surface functionalization and application of Fe_3O_4 magnetic nanoparticles, *Adv. Colloid Interface Sci.* 281 (2020) 102165.

3 Spinel and Inverse Spinel Ferrites for Microwave Absorption

Navdeep Sharma, Sandeep Arya, and Madan Lal

3.1 INTRODUCTION

The growing levels of electromagnetic (EM) radiation have developed a strong concern in our modern era due to its potential human health effects. This radiation is mainly emitted by information technologies, communication equipment, and electronic devices. However, more convenience has been brought about by the widespread utilization of digital devices and rapid advancements in radar technology. It is observed that long exposure to radio frequency (RF) causes unhealthy effects on human beings such as cataracts, skin burns, and hazards stemming from environmental RF field hazards [1, 2]. In light of this, researchers have persistently focused on developing novel materials called "microwave absorbers" that have improved microwave absorption properties. Although microwave absorption phenomena and their mechanisms have long been recognized, collective efforts were carried out by researchers and engineers to explore better solutions [3, 4]. Beyond academia, practical concerns arose from extensive EM wave emissions into human living spaces, causing pressing issues of interference, harmful impact on electronic devices, and health effects. This problem has gained recent attention and relevance in research. Spinel and inverse spinel ferrites have emerged as scientifically interesting and practically valuable solutions to address this demand [5, 6]. Additionally, microwave absorption materials, featuring thinness, lightweight design, broad spectrum absorption, and robust properties, have garnered significant interest for their practical and scientifically grounded role in mitigating electromagnetic interference (EMI) pollution [7, 8]. Adi et al. demonstrated the significance of EMI and the impact of shielding in the electronics field. However, the microwave absorption material features have shown a critical role in EMI reduction. Additionally, it has been noted that high-frequency electronic components frequently experience signal leakage, especially when used in an open environment. But, in closed environments, EMI reflections can lead to energy accumulation and noise interference. Thus, microwave absorbers offer an effective solution to counter EMI effects, ensuring the protection and optimal performance of electronic devices in closed systems [9]. The concept of EMI and shielding was also explored by Wang et al. explaining how disruptive

 DOI: 10.1201/9781032665658-3

electromagnetic energy can be transmitted between electronic devices through radiated, connected pathways, or a combination of both. They also proposed that EMI control could vary based on the coupling or pathway, involving radiation or conduction mechanisms [10]. However, EMI shields operate by absorbing unwanted electromagnetic waves and transforming the energy from these waves into heat, resulting in reducing the interferences and mitigating any disruptive effects. Achieving this effectiveness depends on the tailored design and varies with frequency. Microwave absorbers and radar absorbers safeguard against interference in the microwave and radio frequency bands [11]. Moreover, a material's capacity to absorb electromagnetic waves is determined by its dielectric characteristics, including permittivity and permeability. By introducing proper compositions of dielectric and magnetic materials, it is observed experimentally that the electromagnetic characteristics of these materials have shown more effective results, achieving optimal absorption of electromagnetic energy [12]. Ferrites, including spinel and inverse spinel ferrites, are non-conducting oxides used in microwave absorption to minimize reflections and scattering of microwave signals. They are utilized in industries such as telecommunications, radar, stealth, and EMI shielding as they efficiently absorb and dissipate microwave energy, improving signal quality and reducing interference [13]. Spinel ferrites have a chemical composition represented by the general formula AB_2O_4. They typically consist of 70% Fe_2O_3 (iron oxide) and 30% other metal oxides, like copper oxide (CuO), ferrous oxide (FeO), nickel oxide (NiO), and manganese oxide (MnO) [14]. This formula represents "A" for divalent and "B" for trivalent metal ions. Trivalent ions are typically smaller than divalent ones. The tetrahedral and octahedral positions, when occupied with these ions, have greatly affected the electrical, dielectric, and structural characteristics of spinel ferrites. Hence, these ferrites are divided into three categories: (i) normal spinel, (ii) inverse spinel, and (iii) mixed spinel ferrites. Because these materials have shown low electrical conductivity and minimal dielectric loss, they are highly suitable for high-frequency uses, microwave absorption, and transformer cores [15, 16]. Inverse spinel ferrites typically include copper ferrites [17]. Rabia et al. synthesized spinel ferrites for microwave applications. They analyzed dielectric properties from 0.1 to 20 MHz and measured magnetic properties at 2 KOe. The empirical data analysis showed low dielectric loss and high DC resistivity, highlighting their potential for microwave devices [18]. Further, manganese-based spinel ferrite potential was demonstrated by Hakeem et al. for microwave absorption. Their experimental results demonstrated that, as the concentration of yttrium (Y) increased, specific magnetic properties such as remanence and saturation magnetization reduced, while coercivity value improved. This change in magnetic properties enhanced the material's suitability for absorption applications [19]. Several synthesis methods with different substitution materials have been reported in the literature for spinel and inverse spinel ferrites to meet the microwave absorber efficacy and the fabrication of devices, which depends upon the demand of application [20–23]. This chapter provides an all-inclusive overview to comprehend the microwave absorption fundamentals and EMI solutions based on the spinel and inverse spinel ferrites with a detailed discussion of their mechanism, properties, advantages, and progressive applications.

3.2 FUNDAMENTALS AND MECHANISMS OF MICROWAVE ABSORPTION IN SPINEL AND INVERSE SPINEL FERRITE

The fundamentals of the microwave absorption principle clarify the interaction of EM waves with materials (spinel ferrite/inverse spinel ferrite like $NiZnFe_2O_4$, $MnZnFe_2O_4$, etc.), as shown in Figure 3.1, resulting in separating power into three modes, namely (i) reflection, (ii) absorption, and (iii) transmission. When the incident wave strikes the conductive layer, it triggers numerous internal reflections within the materials. These reflections are the results of materials' heterogeneity, which ultimately leads to dispersion. Basically, the molecular and electronic structure of the material during interaction with the incident microwave causes it to generate heat and facilitate energy absorption. This phenomenon finds diverse applications, from cooking in microwaves to radar and materials science [24].

Additionally, due to the dispersive effect, altering material properties along with multiple reflections enhances the absorption potential of the EM wave absorber. Changing electromagnetic parameters enhances absorption; moreover, the absorbing ferrite materials reduce reflected and transmitted microwaves, helping in EM wave absorption [25]. Fundamentally, microwave absorbers (MAs) have two key electromagnetic properties: (i) dielectric loss and (ii) magnetic loss [26]. In the first case, the reflection loss occurs due to the ferrite material as an absorber and the electric field of the incident electromagnetic radiation, which is overall called dielectric loss [27]. In the second case, when the absorber material interacts with the magnetic field caused by electromagnetic waves, all these interactions are responsible for the magnetic loss [28]. Moreover, two major factors, (i) magnetic permeability represented as $\mu^* = \mu' - j\mu''$ and (ii) electrical permittivity represented as $\varepsilon = \varepsilon' - j\varepsilon''$, explicitly characterize absorber behavior and their performance. However, parameters μ' or ε' correspond to energy storage and μ'' or ε''. The term represents energy

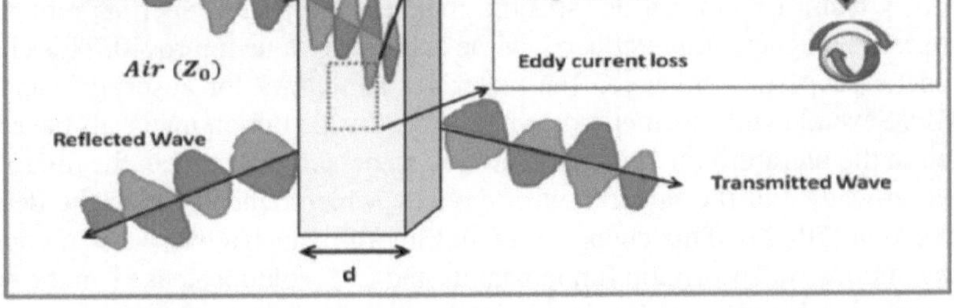

FIGURE 3.1 General pictorial representation of microwave absorption phenomenon.

dissipation by factors such as resonance, relaxation mechanism, and conduction, along with their bipolar behavior [29, 30]. The electromagnetic absorption capacity provides information on reflection loss (RL) measures, which are calculated based on (i) relative complex permittivity and (ii) relative complex permeability factor using the principles of the "transmission line theory." This relationship is represented in Equations (3.1) and (3.2):

$$Z_{in} = z_0 \sqrt{\frac{\mu_r}{\epsilon_r}} \tanh\left[j \left(2\pi fd / c \sqrt{\frac{\epsilon_r}{\mu_r}} \right) \right] \tag{3.1}$$

$$R_L = 20\log \left| \frac{(Z_{in} - Z_0)}{(Z_{in} + Z_0)} \right| \tag{3.2}$$

where the term Z_{in} represents the input impedance of the observer and $Z_0 = \sqrt{\frac{\mu_0}{\varepsilon_0}}$ term represents the impedance of the free space. The term ϵ_r and μ_r, c, f, and d represent the relative complex permittivity and permeability, the velocity of light, the microwave frequency, and the absorber thickness, respectively. When ϵ_r and μ_r values converge, the result is minimal reflection and a higher reflection loss (RL) for the EM wave. However, it is observed that the RL of the EM wave increases when the ϵ_r and μ_r values approach each other, resulting in decreased reflection loss. Effective EM wave absorption necessitates a complementary relationship between dielectric and magnetic losses, resulting in specific impedance characteristics. From Equation (3.1), we can calculate the sample's impedance matching characteristic (Z), as expressed in Equation (3.3):

$$\mathrm{Z} = \left| \frac{Z_{in}}{Z_0} \right| \tag{3.3}$$

Electromagnetic waves experience no reflection on the absorbing material's surface when the impedance (Z) is close to or equal to 1, allowing all incidents to penetrate the material. However, it is experimentally proven that the absolute value of the reflection loss has shown a maximum value, within this impedance approximation value. Conversely, weak reflection loss is obtained, despite good dielectric and magnetic losses, when impedance values are mismatched, because the incident EM waves are either reflected or pass through the front surface of the material without significant dissipation. Therefore, a proper balance between dielectric and magnetic properties is a prime factor for achieving effective EM wave absorption in order to ensure excellent impedance matching. High microwave absorption is thus dependent on maintaining optimal impedance matching [31]. Hence, current research focuses on advanced microwave-absorbing materials: "spinel, and inverse spinel ferrite." These nanomaterials show remarkable efficiency and hold the potential to revolutionize multiple industries.

3.3 SIGNIFICANCE OF THE SPINEL AND INVERSE SPINEL FERRITE IN MICROWAVE ABSORBER RECOGNITION

In general, the peculiar features of the spinel and inverse spinel ferrite to meet the microwave absorber applications in different fields like stealth and space technology and telecommunication, etc. are shown in Figure 3.2.

The word "ferrite" originates from the Latin word "Ferrum" inference "iron." Ferrites are magnetic materials due to their ferromagnetic properties and primarily consist of oxides containing ferric ions. These materials are characterized by their hardness, brittleness, and dark gray or black ceramic appearance [32, 33]. The spinel ferrite is denoted by $MeFe_2O_4$; here, Me represents the general divalent metal ion, for example, Cd, Zn, Cu, Ni, Co, Mn, and Fe are specifically called "divalent transition metals." Spinel describes the crystal structure of "ferrites" featuring a face-centered cubic or cubic close-packed arrangement of oxide ions (O^{2-}). In this lattice, M' cations occupy the $\frac{1}{8}$ th position of the tetrahedral sites, whereas M'' cations fill $\frac{1}{2}$ of the octahedral sites. The respective sites are presented as tetrahedral (for M' cations) and octahedral sites (M'' cations) within the spinel ferrite crystal structure [34, 35]. However, metal ions are disseminated in both "tetrahedral" and "octahedral positions," with Fe^{3+} and Zn^{2+} ions occupying octahedral sites, while ions Mn^{2+}, Ni^{2+} and Fe^{2+} tend to occupy tetrahedral sites. Spinel ferrites exhibit a variety of

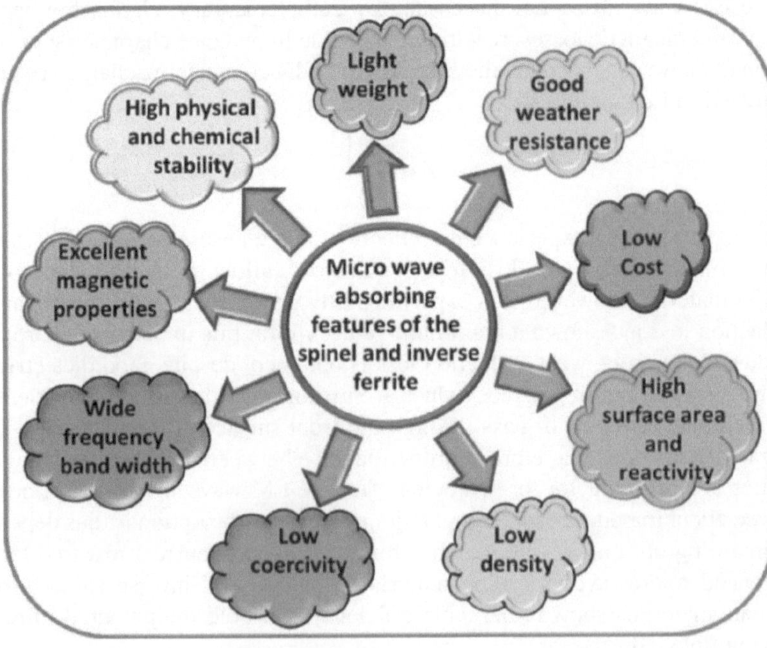

FIGURE 3.2 Properties of spinel and inverse spinel ferrite.

colors, such as green, red, blue, brown, or yellow, though they might be colorless [36, 37]. However, inverse spinel structures are present in MnZn and NiZn ferrites, where a portion of the M (metal ion) atoms, namely, Mn&Ni occupy the tetrahedral sites, while the remaining atoms are specifically Zn positioned at the octahedral sites [38]. Ferrite materials are a class of ceramic compounds characterized by their magnetic oxides and insulating properties. However, in the crystal structure, the arrangement of the metallic ions relative to oxygen ions is responsible for the magnetic behavior in ferrites, but this feature varies among different types of ferrites [39]. Further, spinel and inverse spinel ferrites exhibit low eddy current losses along with high permeability and low permittivity across a broad frequency range. These properties have shown great utility in microwave devices for strong signal coupling and non-reciprocal behavior. They have also shown their applicability in electronics for power generation, cooling, and conversion processes [40, 41]. In addition, high-frequency electronic components like wave absorbers and inductors rely on their initial permeability. However, ferrite materials, including spinels, garnets, and hexa-ferrites, display high-frequency permeability in a tensor format due to their magnetic anisotropy and gyromagnetic characteristics [42–44]. The magnetic dipole moment exhibits precessional motion, which rotates in a specific direction relative to the magnetic bias field, resulting in these properties. But, the reversing direction of the magnetic field causes the dipole moments to rotate in the opposite direction. This unique rotational behavior breaks time-reversal symmetry and leads to non-reciprocal behavior [45]. Basically, for the precessional motion, the optimum value of the frequency required is generated by the strength of the bias – magnetic field, the key factors affecting them are (i) demagnetizing field and (ii) the magneto-crystalline anisotropy fields within the ferrite and the externally applied magnetic field [46]. Furthermore, in ferrite materials, the interaction is maximized between circularly polarized EM waves and dipole moments when the rotation direction of the waves aligns with that of the dipole moments in the materials. This unique interaction makes it possible to create non-reciprocal devices like isolators, circulators, which are important for both testing and operational microwave systems, as the wave's propagation direction affects this interaction. By adjusting the external field, tunable devices like phase shifters and filters can be developed [47]. Spinel and inverse spinel ferrite materials strongly interact with electromagnetic waves at ferromagnetic resonance, resulting in significant wave attenuation. This unique property is utilized for the development of absorber devices for various scientific and technological applications [48, 49]. Thus, advancements in microwave technology are mainly concerned with high-range frequency applications like radar systems and telecommunications required, such as magnetic materials, and the range of frequencies nowadays extends as high as 100 GHz [50]. In polycrystalline ferrites, losses primarily result from defects and anisotropy field variations. Especially in spinel ferrites, Fe^{2+} ions promote hopping conduction. Recent research has focused on understanding these losses and their impact on microwave absorbers, with a focus on their impact on line width. Grain boundaries, the interfaces between crystalline grains, play a vital role in understanding these phenomena. Spinel and inverse spinel ferrites, with their unique magnetic properties and electrical behavior, are crucial for efficient microwave absorption in applications like EMI shielding and radar absorbers [44].

3.3.1 Dielectric Absorbers and Their Losses in Spinel and Inverse Spinel Ferrites

When a material interacts with electromagnetic waves, it experiences dielectric loss due to its inherent dielectric properties. The factor contributing to dielectric loss in a material is the interaction between the material's dipoles and the charge carriers with the incoming electric field, which results in the dissipation of electromagnetic energy as heat [51]. The absorbing material's capacity for dielectric loss arises primarily from two factors: (i) Conductivity loss, which depends on the electrical conductivity, and (ii) polarization loss, encompassing interfacial and atomic polarization respectively, electronic polarization, dipole orientation polarization, and ionic polarization. Among these, atomic and electronic polarization are isolated because they belong to a higher frequency range [52]. Permittivity (ϵ) is a fundamental property that characterizes a material's influence on the electric field during its interaction with electromagnetic waves. It comprises two components: ϵ ' (real part) denotes the material's energy storage capacity, which varies with the electromagnetic wave frequency, while ϵ " (imaginary part) reflects the electric field's attenuation within the medium, revealing its impact on electromagnetic field strength during propagation [53]. The calculation of dielectric loss often comprises the value of the tangent of the dielectric loss angle $\left(\text{tantan}\delta_{\in} = \frac{\epsilon''}{\epsilon'} \right)$ and their value corresponds to the imaginary part of the dielectric constant ($\in$ ") [54]. Qindeel et al. synthesized spinel ferrites with $Mg_{0.6}Li_{0.4}Yb_xFe_{2-x}O_4$ compositions (x = 0.00 and 0.1) for a dielectric study within the frequency range ~0.1 to 20 MHz with different temperature ranges. Experimental analysis revealed that their suitability for the fabrication of microwave devices depends on parameters like dielectric constant values, low dielectric loss, and high DC resistivity [55]. Butt et al. demonstrated the sol–gel method used for the synthesis of $MnLaFe_{2-y}O_4$ a spinel ferrite at different value of "y" given as $(y = 0.00, 0.025, 0.05, 0.075, 0.10)$.The empirical analysis revealed that the "cavity perturbation method" enabled us to find the microwave properties and showed that both real and imaginary parts of the permittivity and permeability decreased with increased frequency. These findings suggest the potential use of this system in microwave absorption applications [56]. Additionally, Sözeri et al. reported the synthesis of spinel ferrites $Ni_xZn_{0.8-x}Mn_{0.1}Fe_2O_4 (0 \leq x \leq 0.8)$. The results showed their excellent absorption properties and the reflection loss reported as −25dBat 10GHz respectively [57]. Singh et al. synthesized $Ni_{0.5}Cu_{0.5-x}Mg_xFe_2O_4$ ferrites and explored their characteristics. The obtained outcomes indicated that these ferrites exhibit a fundamental peak between 10 Hz and 1 KHz. This property, combined with their frequency-dependent resistivity due to electron hopping, makes them valuable as dielectrics and microwave absorbers. The addition of Mg further enhances their performance, especially at low frequencies [58]. Prajna et al. demonstrated EMI shielding with dysprosium (Dy) in X-band and Ku-bands. The results showed that at $x = 0.1$,the maximum shielding effectiveness was obtained and emphasized absorption. LD10FO/CB had 99.6% absorption at 17–18 GHz, with complex permittivity (20–40) and permeability (2–6). Furthermore, it is found that magnetic and electric properties in the case of Dy are enhanced at 4f–3d and 3d–3d coupling systems in the spinel ferrite. These findings suggest the potential of these ceramics as commercial microwave absorbers [59]. Slimani et al. exposed the impact of La^{3+} and Y^{3+} ion substitutions on $Ni_{0.3}Cu_{0.3}Zn_{0.4}Fe_2O_4$nano-spinel ferrites. The results showed that at $x = 0.08$ concentration of LaY, the material exhibits excellent microwave properties and displayed reflection loss of –40dB at 10.5GHz and an absorption bandwidth at10dB. These properties make LaY-doped NiCuZn nano-spinel ferrite a suitable candidate for radar absorption in the X-band [60]. These findings pave the way for future advancements in dielectric absorbers in ferrites and offer promising solutions for improved microwave technology.

3.3.2 Magnetic Absorbers and Their Losses in Spinel and Inverse Spinel Ferrites

Magnetic loss arises when a material encounters an alternating electric field, converting incoming electromagnetic energy into thermal energy through magnetization. Ideal microwave absorbers are highly demanded in electromagnetic absorption applications for their exceptional characteristics. They provide robust absorption, maintain thin profiles, and require minimal filler content along with high broad bandwidth [61]. The ideal wave absorption is achieved by one major point, which is related to infinite reflection loss (R_L = -∞) and typically requires $\mu_r = \epsilon_r$ condition, and it depends on the impedance matching factor related to the material and free space. Furthermore, magnetic materials exhibit low impedance characteristics because they often have high permittivity and moderate permeability [62]. However, the impedance matching factor recognizes the first, while the second is associated with intrinsic electromagnetic (EM) loss. Numerous research surveys have reported that single-dielectric or magnetic lossy materials seldom achieve impedance matching. The solution resides in combining dielectric and magnetic materials to control the frequency-dependent behavior of permittivity factor and permeability factor [63]. Another important result inferred is that the minimum reflection loss or RL dip in the case of ferrite magnetic absorbers is not directly determined by the material's magnetic or dielectric frequency [64, 65]. Further, it has been observed that metals and alloys can exhibit an increase in EM parameters at high frequencies due to their large saturation magnetization [66]. The properties of magnetic material, in the present case it is ferrite, are defined by a complex semi-constant permittivity $\left(\mu r^* = \mu' r - j\mu'' r\right)$. In microwave absorption, the imaginary part $(\mu'' r)$ of complex permeability $\left(\mu r\right)$ plays a pivotal role, resulting in dispersion in frequency regions where $\mu'' r$ predominates while $\mu' r$ remains small [67]. Spinel ferrite with the chemical formulaNiCoFeO$_x$ was synthesized by Pubby et al. Results showed that the doping process enhanced magnetic moment, saturation magnetization, and Curie temperature. Specifically, a high saturation magnetization of $86.12\,emug^{-1}$ and a low coercivity of $694.7\,Oe$ were demonstrated by the composition $Ni_{1-x}Co_xFe_2O_4$ at $x = 0.90$. These characteristics suggested that such a ferrite is a promising candidate for microwave absorber applications, emphasizing the significance of magnetic properties in their interaction with electromagnetic waves [68]. Usman et al. synthesized nickel ferrite nanoparticles co-doped with Mn^{2+} and Cr^{3+}. Magnetic measurements revealed an increase in saturation magnetization of 24.5emu/g and a reduction in coercivity down to 30.3*Oe* showcasing the impact of dopant concentration on magnetic properties [69]. Henrik et al. observed that the magnetic behavior and microwave applicability of the ferrite nanoparticles are directly influenced and depends on the choice of divalent cation and mainly its sharing in octahedral and tetrahedral sites [70]. Xi et al. demonstrated the microwave properties of $CoFe_2O_4/Co_3Fe_7$-Co core–shell structures formed by reducing $CoFe_2O_4$ nanoparticles in an H_2/Ar atmosphere. The results revealed that the improved magnetic and dielectric losses along with effective electromagnetic wave matching attributed to the enhanced performance of microwave absorption [71]. The hydrogen-thermal reduction method was used by Guan et al. to prepare $CoFe_2O_4$/CoFe composites. The results exposed that the carbon-coated composites exhibited dual-band absorption properties in concern with S-C and X [72]. In addition, Golchinvafa et al. reported the synthesized FeCo/$CoFe_2O_4$ composite powders via a solution combustion method. The powders exhibited a porous structure with spherical nanoparticles having 30 nm in size. The results revealed the highest saturation magnetization 99emu/g and observed the lowest coercivity ~450Oe value at $CoFe_2O_4$/17wt%FeCo composite. Additionally, the paraffin--FeCo/$CoFe_2O_4$ composite sample showed its potential for microwave applications, as it was found that their reflection intensity is –20dB in the C band within $4-8$GHz frequency range [73]. Shahab et al. demonstrated the effectiveness of the Co in the Co-Ni-Fe ferrite nanocomposites to escalate their magnetic properties, resulting in coercivity factor (H_c) and magnetization factor (M_s) enhanced in them. The $Co_{0.8}Ni_{0.2}Fe_2O_4$ sample exhibited impressive values, with Ms reaching 93emu/g and Hc at 420Oe. Moreover, these materials demonstrated exceptional microwave absorption, with a significant –26dB reflection loss at 9.7GHz (over 99% absorption) and a wide 4GHz bandwidth in the X-band frequency range [74]. Ghodake et al. revealed in their study that Co^{2+} substitution at the 2.3 to 2.5 GHz range caused an increase in the reflection loss of NiZn ferrite. Results have shown an enhancement of the microwave absorption properties in $Ni-Co-Zn$ ferrites [75]. Rao et al. reported that when cobalt ion is substituted in Ni-Zn ferrite to replace zinc ion, and results show that there is an enhancement in saturation magnetization [76]. Further, Xiang et al. demonstrated that $Co_{0.5}Ni_{0.5}Fe_2O_4$ nanofibres have strong microwave absorption capacities within the Ku band, having a range that lies in between 12and18GHz [77]. Kaur et al. demonstrated that the $Zn-Co-MnZr-Fe_2O_4$ composition of the spinel ferrites showed their utility for wideband microwave absorption range lying between 26.5and40GHz. Further, it was observed that when substitution concentration level MZ0.05toMZ0.25 was given, it caused to reduce coercivity and increase saturation magnetization. In the

Ka −band, MZ0.15 excelled with a −27.99 dB reflection loss and a 6.5 GHz absorption bandwidth (2.2 mm thickness). The simulation technique identified that the composition MZ0.20shows a reflection loss of −42.71 dB. These materials demonstrate efficiency within a range lies between 26.5and 40 GHz and address the 5G technologies developing microwave-absorbing materials [78]. Marwa et al. demonstrated the crystal structure and their specification for $Co_{1-x}Ni_xFe_2O_4$ nano-spinel ferrites for microwave absorptions. Results exposed that a polarization process, namely dipole polarization, and the magnetization reversal processes occurred in the region of inter-resonant processes, resulting in the energy losses (electrical and magnetic losses) due to the reflection. A notable reduction in a reflected wave energy lying in the range −10 to − 21.8 dB creates extensive opportunities for practical applications [79]. Hence, optimizing microwave absorbers requires addressing magnetic loss through impedance matching and combining magnetic/dielectric materials for frequency control.

3.4 EXPLORATION OF THE SPINEL AND INVERSE SPINEL FERRITE MATERIAL: EFFICACY IN MICROWAVE ABSORBER PERFORMANCE

The following section incorporates a detailed discussion of the microwave absorber worth of the spinel and inverse spinel ferrites with (i) carbon-based material, (ii) rare earth, and (iii) transition metals.

3.4.1 Carbon-Based Material

Carbon-based materials are vital for EMI shielding and microwave absorption and help in reducing environmental pollution and health hazards due to their exceptional properties [80]. There is an urgent need to develop carbon-based ferrite materials for microwave absorbers because, with the advancement in telecommunication systems, such as microwave antennae, etc., pollution is a matter of serious alarm. Mahla et al. demonstrated the potential of $CuFe_2O_4$/multi-walled carbon nanotube (MWCNT) composites for microwave absorption under un-annealed and annealed (400°C) conditions. The investigation of the data revealed that reflection loss(RLmin) of −46.4 dB *at* 16 GHz was observed for the un-annealed sample, at a 3.8GHz absorption bandwidth. The annealed sample performed well, with RLmin of −69 dB *at* 10GHz and an 8.5 GHz absorption bandwidth, covering X and Ku bands [81]. Idris et al. demonstrated that a 3D network of multi-walled carbon nanotubes (MWCNTs) and $CoFe_2O_4$within the epoxy matrix. Increasing MWCNT content improved electromagnetic wave absorption because of different thickness values as reported: 1mm, 2mm, 3mm resulting to shift mainly reflection loss peak to lower frequency range. Further, investigation revealed that the largest reflection loss of −28dBat 9GHz was achieved at 2wt% MWCNTs of 3 mm thickness [82]. Ashfaq et al. demonstrated the synthesis of $CoFe_2O_4\ /\ MWCNTs$ composites using the solvothermal technique. The obtained outcomes explained that these composites have shown impressive results: RL_{min} found to be −50.80 dB at a 4.2 mm thickness, with effective absorption bandwidth observed to be 3.36 GHz at a 1.6 mm thickness, outperforming distinctive magnetic composite absorbers' behavior. This investigation highlights the potential of MWCNT-based composites as electromagnetic wave absorbers exhibit (i) highly efficient and (ii) lightweight characteristics [83]. Further, Huang et al. explained the synthesis of the $CoFe_2O_4$ / $CNTs$ composites along with structure escalating their electromagnetic wave absorption merits. Results exposed that the dipole polarization factor increases and the interface losses due to the presence of MWCNTs. The study revealed that the absorption performance factor was found to be best at 25% filling ratio, achieving an RL_{min} of −36.52 dB at 1.9 mm thickness, and a frequency of 3 GHzhas shown the effective absorption bandwidth [84]. Yuan et al. reported the synthesis of nanocomposite materials consisting of $CoFe_2O_4$ and CNTs by the solvothermal method. The final empirical values explained that at a thickness of 1.7 mm, the $CoFe_2O_4$/CNTs composite exhibited a remarkable reflection loss of −37.39 dB, with an effective absorption bandwidth spanning 5.2 GHz, thus recognizing the superiority of the composite toward microwave absorption over the pure cobalt ferrite [85]. Huang et al. exposed in their study that the presence of the MWCNTs, which are wrapped in hollow $CoFe_2O_4$ particles with open holes, resulted in a novel electromagnetic wave absorber. The incorporation of MWCNTs induces dipole polarization, which, when wrapped around $CoFe_2O_4$ particles, enhances the interface loss factor. Analysis revealed that at 25% filling ratio, electromagnetic wave absorption performance was found to be maximum; and this leads to a minimum reflection loss of −36.52 dB at a thickness of 1.9 mm along with 3 GHzeffective absorption bandwidth [86]. Additionally, within the spinel ferrite family, the most promising dielectric material having excellent electromagnetic wave absorber properties is revealed as nickel ferrite ($NiFe_2O_4$). In the experimental observation, it was found that when combined with MWCNTs, it could enhance its EM wave absorption properties by increasing the dielectric loss [87]. Lei Guo et al. demonstrated the synthesis of $NiFe_2O_4$ / $MWCNTs$ with hydrothermal technique to explore their microwave absorption properties. The results revealed that at 11.3 GHz frequency, the optimized sample exhibits the minimum reflection loss (RLmin) value of −19 dB at 1.5 mm thickness and 2.5 GHz bandwidth [88]. Rostami et al. demonstrated the effect of Cu substitution in $Mg - Ni$ spinel ferrite-MWCNT nanocomposites, in order to explore their enhanced EM wave absorption properties. Experimental data analysis showed that the composite achieved an excellent reflection loss value of −39.72 dB at the 1.5 mm thickness, because of effective impedance matching and a balance of dielectric and magnetic loss [89]. Mustaffa et al. demonstrated the effect of MWCNTs when combined with $Ni_{0.5}Zn_{0.5}Fe_2O_4$ in electromagnetic wave absorption features. The results exposed that the prepared composite within a thickness of 3 mm has shown the maximum value of the reflection loss (-19.34 dB)at 8.46 GHz, along with a bandwidth of 1.24 GHz[90]. Chahar et al. demonstrated the synthesis of MWCNT/zinc-doped nickel ferrite nanocomposites using a combination of co-precipitation and physical mixing methods. Their investigation revealed an impressive maximum reflection loss of −51.2 dB *at* 11.2 GHz, with a bandwidth of 0.45 GHz [91]. Dalal et al. exposed the role of MWCNTs with $Ni_{0.3}Zn_{0.4}Co_{0.2}Cu_{0.1}Fe_2O_4$ nanoparticles for microwave absorption. Data analysis showedanexcellentreflectionlossvalueof−25.71 dB at 15.24 GHzwitha1 − mm-thick composite sample [92]. Gao investigated the microwave absorption of $Li-Zn$ ferrite and carbon nanotube composites using a mechanical mixing method to create$LiZn/CNT$. The investigation of the data exposes that at a 5 nm thickness, they achieved the lowest RL of −34 dB and an absorption bandwidth of 5.2 GHz [93]. Shu et al. reported a hybrid composite $rGO/MWCNTs/ZnFe_2O_4$ prepared by the solvothermal approach, creating a 3D conductive network for the microwave absorption features. Results showed that the composite achieved a minimal reflection loss of −22.2 dBand a 2.3 GHzabsorption bandwidth at the sample thickness of 1.0 mm. It was also observed that significant attenuation, impedance matching, magnetic–dielectric synergy, and enhanced value of the conductivity are all included in the absorption mechanism domain [94]. Ding et al. reported the synthesis of copper ferrite/reduced graphene oxide ($CuFe_2O_4/rGO$) using a one-pot hydrothermal method and explored the impact of rGO on the microwave absorption properties. Experimental analysis showed that due to the multi-dipole polarization and the impedance matching factors, composites exhibit a remarkable reflection loss of −58.7 dB at 9.2 GHz and at 2.56 mm thickness and a 5.2 GHz effective bandwidth at 1.85 mm thickness. This study provides insights for designing a highly absorptive material having lightweight, wideband, characteristics [95]. Zhang et al. investigated microwave absorption using rGO decorated with $MnFe_2O_4$. The result revealed that the best microwave absorption properties obtained at 70 wt% filling ratio for the as-prepared hybrid composite (S2) exhibits an RL_{min} value of -47.5 wt% along with 5.2 GHz effective absorption bandwidth at 1.7 mm thickness [96]. Yang et al. synthesized graphene/(1−x)$BaFe_{12}O_{19}$/$CaFe_2O_4$/$x$$CoFe_2O_4$ via the deoxidation method to assess their microwave absorption potential. The outcomes revealed a minimum reflection loss of −36.8 dB, with a 9.9 GHz bandwidth at a 2.9 mm sample thickness [97]. Hence, it is observed that graphene oxide exhibits low density, high surface area, and mechanical strength making it effective for microwave absorption when combined with ferrite materials. This broadens bandwidth, absorbs a wider range of frequencies, enhances impedance matching, and improves microwave attenuation [98, 99]. Their properties explicitly motivate researchers to synthesize carbon-based material with ferrite to enhance microwave absorption for applications like electromagnetic interference shielding and attenuation.

3.4.2 Rare Earth Ion Efficacy:

Sodaee et al. demonstrated the synthesis of Gd^{3+}-doped cobalt ferrites ($CoFe_{2-x}Gd_xO_4$) using a hydrothermal method, with and without annealing. These materials exhibited super-paramagnetic behavior and improved microwave absorption properties, including increased coercivity (H_c) and saturation magnetization (M_s) at low Gd^{3+} concentrations ($x = 0.02$). However, at higher doping levels, these properties decreased. The microwave absorber showed a high magneto-crystalline anisotropy field, making it suitable for covering a wide range of X-band electromagnetic waves [100]. Aman et al. synthesized gadolinium-substituted barium-based spinel ferrites using the sol–gel approach. Results showed that Gd incorporation caused an increase in the activation energy and resistivity. The introduction of Gd led to decreased dielectric constants, reduced DC resistivity at higher temperatures, and correlated with lower remnant and M_s (saturation magnetization), along with increased coercivity (H_c). Furthermore, it was observed that natural magnetic resonance and domain wall pinning might cause frequency dispersion within the range of 0.5–12 GHz during microwave measurements. These findings suggest potential benefits for microwave applications [101]. Leninin et al. investigated $NiGd_xFe_{2-x}O_4$ nanoferrites (x = 0.01 to 0.09) using a sonochemical reaction approach. Results unveiled that adding Gd^{3+} NPs improved dielectric characteristics and induced ferromagnetic behavior at room temperature. Magnetic interaction correlated with lattice strain, and higher Gd^{3+} NP concentrations increased saturation magnetization while decreasing coercivity [102]. Additionally, Neelima et al. reported the microwave-hydrothermal method to examine Dy^{3+}-substituted nickel-zinc ferrites ($Ni_{0.5}Zn_{0.5}Dy_xFe_{2-x}O_4$). The analysis confirmed that at x = 0.07 composition, remarkable microwave absorption properties with a reflection coefficient of −33.24 dBat 10.31 GHz, a 2.59 GHz bandwidth, and losses below −10 dB are exhibited. This suggests their potential as effective X-band microwave absorbers [103]. Munirah et al. reported the synthesis of spinel ferrite nanoparticles $\left[Ni_{0.4}Cu_{0.2}Zn_{0.4}\right]\left(Fe_{2-x}Dy_x\right)O_{4,}x = 0.00 - 0.04)$ using a citrate sol–gel auto-combustion method. Results revealed that increasing Dy^{3+} concentration expanded lattice parameters due to differences in ionic radii between Fe^{3+} and Dy^{3+} ions. In a broad microwave frequency range (1–20 GHz), they observed strong electromagnetic absorption between 1.6 and 2.7 GHz. This absorption was linked to microstructural features, particularly the resonance of domain boundaries [104]. Maria et al. studied Dy^{3+}-substituted soft ferrites $[Mg_{0.5}Cu_{0.25}Co_{0.25}Fe_{2-x}Dy_xo_4],(0.0 \le x \le 0.04)$ sample prepared by sol–gel auto-combustion method and explored their efficacy as the best microwave absorber. The investigation showed that the lowest dielectric loss occurred at x = 0.03 within the frequency range of 8 Hz to 8 MHz. The coercivity (H_c) ranged from 520.82 to 544.02 Oe, and saturation magnetization (M_s) ranged from 20.5841 to 21.1473 emu/g [105]. Kumbhar et al. studied rare earth ion (Sm^{3+}, Dy^{3+}) substitution in $MgFe_2O_4$ prepared via chemical combustion. The investigation showed higher reflection coefficients compared to $MgFe_2O_4$. Initial permeability increased with rare earth substitution at x = 0.15, then decreased at higher substitutions. Saturation magnetization showed anomalous behavior and potential use in microwave absorption materials [106]. Majid et al. reported the Sm-doped Co ferrite with composition $\left(Co_{1+x}Sm_xFe_{2-2x}O_4\right)$ at different values:$x = 0.00,\ 0.03,\ and\ 0.09$ with the help of sol–gel auto-combustion method. The investigation exposed that at $x = 0.09$ composition, the corresponding nonlinear coercivity increases and exhibits the highest dielectric constant value. Results also showed improved absorptivity along with multiple absorption peaks, which is attributed to the strong LS coupling of the cations and superexchange interaction factor [107]. Stergiou et al. reported the rare earth-substituted cubic ferrites for microwave absorption $Ni - Co\ and\ Ni - Co - Zn)$. Samples $\left(R = Y\ and\ La, x = 0,0.02\right)$were prepared with compositions $Ni_{0.5}Co_{0.5}Fe_{2-x}R_xO_4$ *and* $Ni_{0.25}Co_{0.5}Zn_{0.25}Fe_{2-x}R_xO_4$ using solid-state reactions. The ferrites displayed narrowband return losses near ferromagnetic resonance. Rare earth doping shifted this effect to lower frequencies, resulting in high RL peaks at 4.1 GHz and 5GHz for $Ni-Co-Zn$ and 18 GHz and 19 GHz for$Ni-Co$. This suggests the potential for suppressing electromagnetic reflections well above 1GHz [108]. In addition, Srinivas et al. reported rare earth-doped (RE) Mn-Ni-Zn ferrite materials ($Mn_{0.1}Ni_{0.7}Zn_{0.2}RE_{0.5}Fe_{1.5}O_4, RE = Dy, Ce, Sm, Er$) via the sol–gel auto-combustion method. This composition exhibited the highest saturation magnetization (41.2 emu/g). In terms of complex permittivity, ε ' increased with frequency (8.2 GHz: 5.421–5.856, 12.4 GHz: 5.538–6.066), and ε'' displayed oscillatory behavior (8.2 GHz: 0.092–0.104, 12.4 GHz: 0.042–0.105). For complex permeability, μ ' showed a wavy pattern and μ '' decreased with frequency, with weak resonance peaks around 9–10 GHz. μ ' values ranged from 0.869 to 0.958 at 8.2 GHz and 0.903 *to* 0.925 *at* 12.4 GHz, while μ '' values ranged from 0.269 *to* 0.334 *at* 8.2 GHz *and* 0.059 *to* 0.068 *at* 12.4 GHz. These findings suggest the potential of this composition as a microwave absorber in the $X-band$ frequency due to its microwave behavior [109]. Hence, explicitly clear that the rare earth ion doping enhances ferrite materials for efficient microwave absorption, promising advancements in electromagnetic wave control and communication technologies.

3.4.3 Transitional Metal Efficacy

Mandal et al. studied the microwave absorption properties of metal ferrite nano-hollow spheres (NHS) in an epoxy matrix across 1–20 GHz. Results showed that substituting M^{2+} ions $(M = Mn, Co, Ni, Cu, Zn)$ enhanced absorption. $MnFe_2O_4$ NHS demonstrated an optimal RL of (32.7 dB), total shielding (42 dB) and a high attenuation constant (α) in (196) nepers per meter (Np/m). RL peaks at 4GHz and 12 GHz resulted from magnetic resonance and dielectric relaxation, with an additional peak at 19.3 GHz due to impedance matching. $MnFe_2O_4$NHS composites at 4.2 mm thickness (50 wt%) achieved a maximum RL of 45.6 dBand a broad effective bandwidth (<10 dB) of 3.6 GHz [110]. Tong et al. synthesized $Co_{1-x}Ni_xFe_2O_4 (x = 0, 0.1, 0.3, 0.5)$ ferrite materials via sol–gel combustion. Analysis of the results exposed that by substituting Ni^{2+} for Co^{2+} caused lattice distortion, altering magnetic properties. Smaller grain sizes resulted from Ni doping, affecting space charge polarization and introducing electron polarization. The Ni doping amount of 0.1 for the $Co_{1-x}Ni_xFe_2O_4$ composition achieved a reflection loss of 37.66 dB *at* 16.24GHz with 2.1 mm thickness and an effective bandwidth of 2.64 GHz, making it suitable for lightweight, thin, and wideband microwave absorption material [111].

Mandeep et al. synthesized Co-substituted zinc spinel ferrites ($Zn_{1-x}Co_xFe_2O_4 with\ x = 0.00, 0.25, 0.50, 0.75,$ *and* 1.00) with varying Co concentrations using a sol–gel citrate route and sintered the materials at 1000°C for 6 h. The empirical consequences show that all compositions achieved over 90% absorption in the frequency range of $18-26.5$ GHz, indicating their potential for effective microwave absorption at higher frequencies [112]. Additionally, Ünal et

TABLE 3.1
Overall View on Transition Metal-Based Ferrite

Ferrite Material	Synthesis Method	Frequency (GHz)	RL (dB)	Thickness (nm)	Bandwidth	Reference
$Mn_{1-x}Zn_xFe_2O_4$	Sol–gel	11.05	–42.06	3.0	3.34	[116]
$FeCo/CoFe_2O_4$	Liquefied phase reduction	14.8	–34.8	1.5	5.1	[117]
$ZnFe_2O_4$/MWCNTs	Solvothermal method	13.4	–55.5	1.5	3.6	[118]
$NiFe_2O_4$ /MWCNTs	Solvothermal method	11.8	–42.3	1.2	3.8	[119]
$FeCo/CoFe_2O_4$	Combustion method	17.0	–20	4.0	–	[120]
$Ni_{0.6}Zn_{0.4}(MgZr)_xFe_{2-2x}O_4$	Sol–gel	10.55	–35.10	4.0	2.8	[121]
$Ni_{1-x}Ce_xFe_{2-x}Zn_xO_4$	Auto-combustion method	15.52	–21.44	2.5	5.44	[122]
FeSiAl/NiZn ferrite	Co-precipitation	10.7	–29.2	2.5	4	[123]
$Ni_{0.6}Zn_{0.4}(MgZr)_xFe_{2-2x}O_4$	Sol–gel	17.21	–33.4	1.6	18	[124]
$Ni_{0.6}Zn_{0.4}(MgZr)_xFe_{2-2x}O_4$	Sol–gel	10.552	–35.10	4	–	[125]
$Zn_{1-x}Co_xFe_2O_4$	Sol–gel	12.4	–10	2.2	5.6	[126]
$Mn_{1-x}Zr_xFe_2O_4$	Sol–gel	$3.1x10^{-4}$	–14	4.72	–	[127]

al. reported $Mg-Ti$ substituted $Ni-Zn$ spinel ferrites $Ni_{0.5}Zn_{0.5}Mg_xTi_xFe_{2-2x}O_4$ using the solid-state reaction method. Results indicated that the sample composition $Ni_{0.5}Zn_{0.5}Mg_{0.2}Ti_{0.2}Fe_{1.6}O_4$ exhibited outstanding microwave properties with a reflection loss of approximately -29 *dB* at 10.4 *GHz* for a 3 *mm* thickness, and better than -35 *dB* at 16 *GHz* with a reduced thickness of 2 mm. This material is promising as a microwave absorber in both X- and Ku-bands [113]. Neha et al. investigated Co and Zr substitutions in $Ni-Zn$ spinel ferrites $\left(Ni_{0.6}Zn_{0.4}\left(CoZr\right)xFe_{2-2x}O_4\right)$ using a sol–gel citrate route. They found substantial dielectric and magnetic losses, with remarkable minimum reflection losses of $-44.92\ dB\left(x=0.20\right)$ and $-40.43\ dB\left(x=0.25\right)$ *at* 4.25 *mm* thickness in the $X-$ band. These materials hold great promise as efficient microwave absorbers in this frequency range [114]. Gao et al. investigated $Li_{0.25+x/3}Zn_{0.5}Fe_{2.25-x}Ti_{2x/3}O_4$ ferrites synthesized via the sol–gel method. The results indicated that the dielectric constant increases with higher Li and Ti contents, attributed to the higher concentration of Fe^{2+} ions. A 5-mm-thick Li-Zn ferrite demonstrated superior absorption properties compared to variations in thickness, showing promise as a potential microwave absorber [115].

Table 3.1 explicitly showed more in-depth on the microwave absorbance compatibility of the transition metals with spinel and inverse spinel ferrites to enhance their applicability. This analysis reveals the complex connections between material composition, synthesis methods, thickness, frequency, bandwidth, and electromagnetic performance. Hence, this is essential for optimizing spinel and inverse spinel ferrite materials in high-frequency applications to improve their capacity for absorbing and reducing EM waves.

3.5 CONCLUSION

In conclusion, the synergetic combination of the spinel and inverse spinel ferrite materials with additives, as discussed in this chapter, like carbon-based substances, rare earth ions, and transition metals, has demonstrated exceptional microwave absorption properties. This research field's future entails the further refinement of material compositions and processing methods to enhance microwave absorption and potentially yield advanced materials for electromagnetic wave management in radar, telecommunications, and defense applications. The ongoing focus on lightweight, wideband, and efficient materials holds great promise for shaping the future of electromagnetic wave technology and its multifaceted applications.

REFERENCES

1. Sassi, O., Naseef, M., Hervé, P., & Kadi, M. (2020, September). Evaluation method of wireless communication system performance based on PER-function coupled with full wave simulation in presence of EM-interference. In *2020 International symposium on electromagnetic compatibility-EMC Europe* (pp. 1–6). IEEE.
2. Dehghan, N., & Taeb, S. (2013). Adverse health effects of occupational exposure to radiofrequency radiation in airport surveillance radar operators. *Indian Journal of Occupational and Environmental Medicine*, 17(1), 7.

3. Zeng, X., Cheng, X., Yu, R., & Stucky, G. D. (2020). Electromagnetic microwave absorption theory and recent achievements in microwave absorbers. *Carbon*, 168, 606–623.
4. Gul, I. H., & Maqsood, A. (2008). Structural, magnetic and electrical properties of cobalt ferrites prepared by the sol–gel route. *Journal of Alloys and Compounds*, 465(1–2), 227–231.
5. Amiri, M., Salavati-Niasari, M., & Akbari, A. (2019). Magnetic nanocarriers: Evolution of spinel ferrites for medical applications. *Advances in Colloid and Interface Science*, 265, 29–44.
6. Choi, J., Kim, D., Hong, S. J., Zhang, X., Hong, H., Chun, H., Han, B., Lee, L. Y. S., & Piao, Y. (2022). Tuning the electronic structure and inverse spinel degree of inverse spinel ferrites by integrating samarium orthoferrite for efficient water oxidation. *Applied Catalysis B: Environmental*, 315, 121504.
7. Green, M., & Chen, X. (2019). Recent progress of nanomaterials for microwave absorption. *Journal of Materiomics*, 5(4), 503–541.
8. Yadav, R. S., Jamatia, T., Kuřitka, I., Vilčáková, J., Škoda, D., Urbánek, P., & Havlica, J. (2020). Excellent, lightweight and flexible electromagnetic interference shielding nanocomposites based on polypropylene with MnFe2O4 spinel ferrite nanoparticles and reduced graphene oxide. *Nanomaterials*, 10(12), 2481.
9. Adi, W. A., Yunasfi, Y., Mashadi, M., Winatapura, D. S., Mulyawan, A., Sarwanto, Y., Gunanto, Y. E., & Taryana, Y. (2019). Metamaterial: Smart magnetic material for microwave absorbing material. *Electromagnetic Fields and Waves,* 1–18.
10. Wang, C., Murugadoss, V., Kong, J., He, Z., Mai, X., Shao, Q., Chen, Y., et al. (2018). Overview of carbon nanostructures and nanocomposites for electromagnetic wave shielding. *Carbon,* 140, 696–733.
11. Ren, F., Shi, Y., Ren, P., Si, X., & Wang, H. (2017). Cyanate ester resin filled with graphene nanosheets and $NiFe_2O_4$–reduced graphene oxide nanohybrids for efficient electromagnetic interference shielding. *Nano*, 12(6), 1750066.
12. Liu, C., Yu, D., Kirk, D. W., & Xu, Y. (2017). Electromagnetic wave absorption of silicon carbide based materials. *RSC Advances,* 7(2), 595–605.
13. Chakradhary, V. K., & Akhtar, M. J. (2021). Absorption properties of CNF mixed cobalt nickel ferrite nanocomposite for radar and stealth applications. *Journal of Magnetism and Magnetic Materials*, 525, 167592.
14. Zahi, S., Hashim, M., & Daud, A. R. (2007). Synthesis, magnetic properties and microstructure of Ni–Zn ferrite by sol–gel technique. *Journal of Magnetism and Magnetic Materials*, 308(2), 177–182.
15. Hashim, M., Kumar, S., Shirsath, S. E., Kotnala, R. K., Shah, J., & Kumar, R. (2013). Synthesis and characterizations of Ni2+ substituted cobalt ferrite nanoparticles. *Materials Chemistry and Physics*, 139(2–3), 364–374.
16. Tahir Farid, H. M., Ahmad, I., Ali, I., Mahmood, A., & Ramay, S. M. (2018). Structural and dielectric properties of copper-based spinel ferrites. *The European Physical Journal Plus*, 133, 1–12.
17. Shah, M. S., Ali, K., Ali, I., Mahmood, A., Ramay, S. M., & Farid, M. T. (2018). Structural and magnetic properties of praseodymium substituted barium-based spinel ferrites. *Materials Research Bulletin*, 98, 77–82.
18. Qindeel, R., Alonizan, N. H., Alghamdi, E. A., & Awad, M. A. (2021). Synthesis and characterization of spinel ferrites for microwave devices. *Journal of Sol-Gel Science and Technology*, 97, 593–599.
19. Hakeem, A., Alshahrani, T., Farid, H. M. T., Khan, A. R., Alhossainy, M. H., Laref, A., & Ali, I. (2021). Manganese-based spinel ferrites for microwave absorption. *Journal of Materials Science: Materials in Electronics*, 32, 2557–2563.

20. Masrour, R., Hamedoun, M., & Benyoussef, A. (2010). Magnetic properties of B and AB-spinels Znl− xMxFe2O4 (M= Ni, Mg) materials. *Journal of Alloys and Compounds*, 503(2), 299–302.
21. Song, Q., & Zhang, Z. J. (2004). Shape control and associated magnetic properties of spinel cobalt ferrite nanocrystals. *Journal of the American Chemical Society*, 126(19), 6164–6168.
22. Kaur, M., & Bahel, S. (2023). Fabrication of $Zn_{0.25}Co_{0.75}(MnZr)xFe_{2-2x}O_4$ spinel ferrites as broadband microwave absorbers in 26.5–40 GHz frequency range. *Journal of Alloys and Compounds*, 954, 170154.
23. Mangalaraja, R. V., Ananthakumar, S., Manohar, P., & Gnanam, F. D. (2003). Initial permeability studies of Ni–Zn ferrites prepared by flash combustion technique. *Materials Science and Engineering: A*, 355(1–2), 320–324.
24. Zeng, X., Cheng, X., Yu, R., & Stucky, G. D. (2020). Electromagnetic microwave absorption theory and recent achievements in microwave absorbers. *Carbon,* 168, 606–623.
25. Pang, H., Duan, Y., Huang, L., Song, L., Liu, J., Zhang, T., ... Liu, X. (2021). Research advances in composition, structure and mechanisms of microwave absorbing materials. *Composites Part B: Engineering*, 224, 109173.
26. Pinto, S. S., & Rezende, M. C. (2018). Performance prediction of microwave absorbers based on POMA/carbon black composites in the frequency range of 8.2 to 20 GHz. *Journal of Aerospace Technology and Management,* 10. https://doi.org/10.5028/jatm.v10.898.
27. Nelson, S. O. (1999). Dielectric properties measurement techniques and applications. *Transactions of the ASAE*, 42(2), 523–529.
28. Chen, L.-F., Ong, C. K., Neo, C. P., Varadan, V. V., & Varadan, V. K. (2004). *Microwave electronics: Measurement and materials characterization.* John Wiley & Sons.
29. Widanarto, W., Ekaputra, A. I., Effendi, M., Cahyanto, W. T., Ghoshal, S. K., Kurniawan, C., Handoko, E., & Alaydrus, M. (2020). Neodymium ions activated barium ferrite composites for microwave X-band absorber applications: Synthesis and characterizations. *Composites Communications,* 19, 51–55. https://doi.org/10.1016/j.coco.2020.03.006.
30. Mensah, E. E., Abbas, Z., Azis, R. S., & Khamis, A. M. (2022). Effect of microstructure on complex permittivity and microwave absorption properties of recycled α-Fe_2O_3 nanopowder prepared by high-energy ball milling technique. *Materials Express,* 12(2), 319–326.
31. Liu, Z., Bai, G., Huang, Y., Li, F., Ma, Y., Guo, T., He, X., Lin, X., Gao, H., & Chen, Y. (2007). Microwave absorption of single-walled carbon nanotubes/soluble cross-linked polyurethane composites. *The Journal of Physical Chemistry C,* 111(37), 13696–13700.
32. Vollath, D. (2008). Nanomaterials: An introduction to synthesis, properties, and application. *Environmental Engineering and Management Journal,* 7(6), 865–870.
33. Elmahaishi, M. F., Azis, R. S., Ismail, I., Mustaffa, M. S., Abbas, Z., Matori, K. A., Muhammad, F. D., et al. (2021). Structural, electromagnetic, and microwave properties of magnetite extracted from mill scale waste via conventional ball milling and mechanical alloying techniques. *Materials,* 14(22), 7075.
34. Goldman, A. (2006). *Modern ferrite technology.* Springer Science & Business Media.
35. Chatterjee, J., Haik, Y., & Chen, C.-J. (2003). Size dependent magnetic properties of iron oxide nanoparticles. *Journal of Magnetism and Magnetic Materials*, 257(1), 113–118.
36. Smart, L. E., & Moore, E. A. (2012). *Solid state chemistry: An introduction.* CRC Press.

37. Srivastava, R., & Yadav, B. C. (2012). Ferrite materials: Introduction, synthesis techniques, and applications as sensors. *International Journal of Green Nanotechnology*, 4(2), 141–154.
38. Haque, M. M., Huq, M., & Hakim, M. A. (2005). Structural characterization and complex magnetic permeability spectra of Mn-Zn ferrites. *Journal of Applied Science and Technology*, 4(1), 11–15.
39. Kadyrzhanov, K. K., Egizbek, K., Kozlovskiy, A. L., & Zdorovets, M. V. (2019). Synthesis and properties of ferrite-based nanoparticles. *Nanomaterials*, 9(8), 1079. https://doi.org/10.3390/nano9081079
40. Mallégol, S., Quéffélec, P., & Le Floc'h, M. (2003). Permeability tensor of magnetized ferrites at microwave frequencies: A comparison between theory and experiment. *Journal of Applied Physics*, 93(10), 7471–7473.
41. Qiao, J., Zhang, X., Xu, D., Kong, L., Lv, L., Yang, F., Wang, F., Liu, W., & Liu, J. (2020). Design and synthesis of TiO_2/Co/carbon nanofibers with tunable and efficient electromagnetic absorption. *Chemical Engineering Journal*, 380, 122591. https://doi.org/10.1016/j.cej.2019.122591
42. Tsutaoka, T., Kasagi, T., Nakamura, T., & Hatakeyama, K. (1997). High frequency permeability of Mn-Zn ferrite and its composite materials. *Le Journal de Physique IV*, 7(C1), C1–557.
43. Landau, L., & Lifshitz, E. (1992). On the theory of the dispersion of magnetic permeability in ferromagnetic bodies. In *Perspectives in theoretical physics* (pp. 51–65). Pergamon.
44. Harris, V. G., Geiler, A., Chen, Y., Yoon, S. D., Wu, M., Yang, A., Chen, Z., et al. (2009). Recent advances in processing and applications of microwave ferrites. *Journal of Magnetism and Magnetic Materials*, 321(14), 2035–2047. https://doi.org/10.1016/j.jmmm.2009.02.090
45. Kirilyuk, A., Kimel, A. V., & Rasing, T. (2010). Ultrafast optical manipulation of magnetic order. *Reviews of Modern Physics*, 82(3), 2731. https://doi.org/10.1103/RevModPhys.82.2731
46. Özgür, Ü., Alivov, Y., & Morkoç, H. (2009). Microwave ferrites, part 1: Fundamental properties. *Journal of Materials Science: Materials in Electronics*, 20, 789–834. https://doi.org/10.1007/s10854-009-9923-7
47. Nagulu, A., Reiskarimian, N., & Krishnaswamy, H. (2020). Non-reciprocal electronics based on temporal modulation. *Nature Electronics*, 3(5), 241–250. https://doi.org/10.1038/s41928-020-0419-4
48. Trukhanov, S. V., Trukhanov, A. V., Kostishyn, V. G., Panina, L. V., Trukhanov, A. V., Turchenko, V. A., Tishkevich, D. I., et al. (2017). Effect of gallium doping on electromagnetic properties of barium hexaferrite. *Journal of Physics and Chemistry of Solids*, 111, 142–152. https://doi.org/10.1016/j.jpcs.2017.08.003
49. Wahaab, F. A., Adebayo, L. L., Adekoya, A. A., Hakeem, I. G., Alqasem, B., & Obalalu, A. M. (2020). Physiochemical properties and electromagnetic wave absorption performance of $Ni_{0\cdot5}Cu_{0\cdot5}Fe_2O_4$ nanoparticles at X-band frequency. *Journal of Alloys and Compounds*, 836, 155272. https://doi.org/10.1016/j.jallcom.2020.155272
50. Gross, M. C., Callahan, P. T., Clark, T. R., Novak, D., Waterhouse, R. B., & Dennis, M. L. (2010). Tunable millimeter-wave frequency synthesis up to 100 GHz by dual-wavelength Brillouin fiber laser. *Optics Express*, 18(13), 13321–13330. https://doi.org/10.1364/OE.18.013321
51. Wang, B., Wu, Q., Fu, Y., & Liu, T. (2021). A review on carbon/magnetic metal composites for microwave absorption. *Journal of Materials Science & Technology*, 86, 91–109. https://doi.org/10.1016/j.jmst.2020.12.040

52. Wang, H., & Ma, H. (2020). Enhancing the microwave absorbing properties of molybdenum dioxide by designing a double-layered structure. *Materials Research Bulletin,* 122, 110692. https://doi.org/10.1016/j.materresbull.2019.110692
53. Cui, L., Han, X., Wang, F., Zhao, H., & Du, Y. (2021). A review on recent advances in carbon-based dielectric systems for microwave absorption. *Journal of Materials Science,* 56, 10782–10811. https://doi.org/10.1007/s10853-021-05949-y
54. Kuang, J., Xiao, T., Zheng, Q., Xiong, S., Wang, Q., Jiang, P., Liu, W., & Cao, W. (2020). Dielectric permittivity and microwave absorption properties of transition metal Ni and Mn doped SiC nanowires. *Ceramics International,* 46(9), 12996–13002. https://doi.org/10.1016/j.ceramint.2020.01.165
55. Qindeel, R., Alonizan, N. H., Alghamdi, E. A., & Awad, M. A. (2021). Synthesis and characterization of spinel ferrites for microwave devices. *Journal of Sol-Gel Science and Technology,* 97, 593–599.
56. Butt, K. Y., Aman, S., AlObaid, A. A., Al-Muhimeed, T. I., Rehman, A., Hegazy, H. H., ... Farid, H. M. T. (2021). The study of structural, magnetic and dielectric properties of spinel ferrites for microwave absorption applications. *Applied Physics A,* 127(9), 714.
57. Sözeri, H., Genç, F., Ünal, B., Baykal, A., & Aktaş, B. (2016). Magnetic, electrical and microwave properties of Mn–Co substituted NixZn0, 8-xFe2O4 nanoparticles. *Journal of Alloys and Compounds,* 660, 324–335.
58. Singh, N., Agarwal, A., Sanghi, S., & Khasa, S. (2012). Dielectric loss, conductivity relaxation process and magnetic properties of Mg substituted Ni–Cu ferrites. *Journal of Magnetism and Magnetic Materials,* 324(16), 2506–2511.
59. Mohapatra, P. P., & Dobbidi, P. (2023). Development of spinel ferrite-based composites for efficient EMI shielding. *Materials Chemistry and Physics,* 301, 127581.
60. Almessiere, M. A., Slimani, Y., Korkmaz, A. D., Baykal, A., Güngüneş, H., Sözeri, H., ... Manikandan, A. (2019). Impact of La 3+ and Y 3+ ion substitutions on structural, magnetic and microwave properties of Ni 0.3 Cu 0.3 Zn 0.4 Fe 2 O 4 nanospinel ferrites synthesized via sonochemical route. *RSC Advances,* 9(53), 30671–30684.
61. Shu, R., Wu, Y., Li, W., Zhang, J., Liu, Y., Shi, J., & Zheng, M. (2020). Fabrication of ferroferric oxide–carbon/reduced graphene oxide nanocomposites derived from Fe-based metal–organic frameworks for microwave absorption. *Composites Science and Technology,* 196, 108240. https://doi.org/10.1016/j.compscitech.2020.108240
62. Elmahaishi, M. F., Azis, R. S., Ismail, I., Mustaffa, M. S., Abbas, Z., Matori, K. A., Muhammad, F. D., et al. (2021). Structural, electromagnetic, and microwave properties of magnetite extracted from mill scale waste via conventional ball milling and mechanical alloying techniques. *Materials,* 14(22), 7075. https://doi.org/10.3390/ma14227075
63. Wang, L., Li, X., Shi, X., Huang, M., Li, X., Zeng, Q., & Che, R. (2021). Recent progress of microwave absorption microspheres by magnetic–dielectric synergy. *Nanoscale,* 13(4), 2136–2156. https://doi.org/10.1039/D0NR07849E
64. Nazlan, R., Ismail, I., Azis, R. S., Abbas, Z., Ibrahim, I. R., Idris, F. M., Shafiee, F. N., Aripin, A. S., & Busra, N. A. N. (2018). Dependence of magnetic and microwave loss on evolving microstructure in yttrium iron garnet. *Journal of Materials Science: Materials in Electronics,* 29, 8688–8700. https://doi.org/10.1007/s10854-018-9007-6
65. Shi, Y., Yin, Y., Zhang, Y., Hu, Y., & Liu, W. (2019). Preparation and microwave absorption properties of C@Fe_3O_4 magnetic composite microspheres. *Materials,* 12(15), 2404. https://doi.org/10.3390/ma12152404
66. Yang, Y., Zhang, B. S., Xu, W. D., Shi, Y. B., Zhou, N. S., & Lu, H. X. (2004). Preparation and electromagnetic characteristics of a novel iron-coated carbon fiber. *Journal of Alloys and Compounds,* 365(1–2), 300–302. https://doi.org/10.1016/j.jallcom.2003.10.072

67. Cuellar, C., Tan, W., Margueron, X., Benabou, A., & Idir, N. (2012, May). Measurement method of the complex magnetic permeability of ferrites in high frequency. In *2012 IEEE international instrumentation and measurement technology conference proceedings* (pp. 63–68). IEEE.
68. Pubby, K., Babu, K. V., & Narang, S. B. (2020). Magnetic, elastic, dielectric, microwave absorption and optical characterization of cobalt-substituted nickel spinel ferrites. *Materials Science and Engineering: B*, 255, 114513.
69. Ahmad, U., Afzia, M., Shah, F., Ismail, B., Rahim, A., & Khan, R. A. (2022). Improved magnetic and electrical properties of transition metal doped nickel spinel ferrite nanoparticles for prospective applications. *Materials Science in Semiconductor Processing*, 148, 106830.
70. Andersen, H. L., Saura-Múzquiz, M., Granados-Miralles, C., Canévet, E., Lock, N., & Christensen, M. (2018). Crystalline and magnetic structure–property relationship in spinel ferrite nanoparticles. *Nanoscale*, 10(31), 14902–14914.
71. Xi, L., Wang, Z., Zuo, Y., & Shi, X. (2010). The enhanced microwave absorption property of CoFe2O4nanoparticles coated with a Co3Fe7–Co nanoshell by thermal reduction. *Nanotechnology*, 22, 045707.
72. Guan, Z. J., Jiang, J. T., Chen, N., Gong, Y. X., & Zhen, L. (2018). Carbon-coated CoFe–CoFe2O4 composite particles with high and dual-band electromagnetic wave absorbing properties. *Nanotechnology*, 29(30), 305604.
73. Golchinvafa, S., Masoudpanah, S. M., & Jazirehpour, M. (2019). Magnetic and microwave absorption properties of FeCo/CoFe2O4 composite powders. *Journal of Alloys and Compounds*, 809, 151746.
74. Torkian, S., Ghasemi, A., & Razavi, R. S. (2017). Cation distribution and magnetic analysis of wideband microwave absorptive CoxNi1– xFe2O4 ferrites. *Ceramics International*, 43(9), 6987–6995.
75. Ghodake, J. S., Kambale, R. C., Shinde, T. J., Maskar, P. K., & Suryavanshi, S. S. (2016). Magnetic and microwave absorbing properties of Co2+ substituted nickel–zinc ferrites with the emphasis on initial permeability studies. *Journal of Magnetism and Magnetic Materials*, 401, 938–942.
76. Rao, B. P., & Caltun, O. F. (2006). Microstructure and magnetic behaviour of Ni-Zn-Co ferrites. *Journal of Optoelectronics and Advanced Materials*, 8(3), 995.
77. Xiang, J., Chu, Y., Zhang, X., & Shen, X. (2012). Magnetic and microwave absorption properties of electrospun Co0. 5Ni0. 5Fe2O4 nanofibers. *Applied Surface Science*, 263, 320–325.
78. Kaur, M., & Bahel, S. (2023). Fabrication of Zn0. 25Co0. 75 (MnZr) xFe2–2xO4 spinel ferrites as broadband microwave absorbers in 26.5–40 GHz frequency range. *Journal of Alloys and Compounds*, 954, 170154.
79. Hussein, M. M., Saafan, S. A., Abosheiasha, H. F., Zhou, D., Klygach, D. S., Vakhitov, M. G., Trukhanov, S. V., et al. (2023). Crystal structure and peculiarities of microwave parameters of $Co_{1-x}Ni_xFe_2O_4$ nano spinel ferrites. *RSC Advances,* 13(38), 26879–26891.
80. Ibrahim Lakin, I., Abbas, Z., Azis, R. S., Ibrahim, N. A., & Abd Rahman, M. A. (2020). The effect of MWCNTs filler on the absorbing properties of OPEFB/PLA composites using Microstrip line at microwave frequency. *Materials*, 13(20), 4581.
81. Tahamipoor, M., & Hekmatara, H. (2023). Superior microwave absorption ability of CuFe 2 O 4/MWCNT at whole Ku-band and half X-band. *Physical Chemistry Chemical Physics*, 25(18), 13145–13154.
82. Idris, F. M., Kaco, H., Mohd, S. M., Jan, N. M., Esa, F., & Idris, Z. M. (2023). Electromagnetic wave reduction of multiwalled carbon nanotubes (MWCNT) mixed nanometer CoFe2O4 at higher frequency range. *Materials Today: Proceedings*, 74, 462–470.

83. Ashfaq, M. Z., Ashfaq, A., Majeed, M. K., Saleem, A., Wang, S., Ahmad, M., ... Gong, H. (2022). Confined tailoring of CoFe2O4/MWCNTs hybrid-architectures to tune electromagnetic parameters and microwave absorption with broadened bandwidth. *Ceramics International*, 48(7), 9569–9578.
84. Huang, D., Dai, J., Wang, Q., Liu, H., & Li, Z. (2023). MWCNTs wrapped in hollow with open holes of CoFe2O4 as high-performance electromagnetic wave absorbers. *Journal of Alloys and Compounds*, 944, 169194.
85. Yuan, Y., Wei, S., Liang, Y., Wang, B., Wang, Y., Xin, W., ... Zhang, Y. (2021). Solvothermal assisted synthesis of CoFe2O4/CNTs nanocomposite and their enhanced microwave absorbing properties. *Journal of Alloys and Compounds*, 867, 159040.
86. Huang, D., Dai, J., Wang, Q., Liu, H., & Li, Z. (2023). MWCNTs wrapped in hollow with open holes of CoFe2O4 as high-performance electromagnetic wave absorbers. *Journal of Alloys and Compounds*, 944, 169194.
87. Soomro, S. A., Gul, I. H., Khan, M. Z., Naseer, H., & Khan, A. N. (2017). Dielectric properties evaluation of NiFe2O4/MWCNTs nanohybrid for microwave applications prepared via novel one step synthesis. *Ceramics International*, 43(5), 4090–4095.
88. Guo, L., He, Y., Chen, D., Du, B., Cao, W., Lv, Y., & Ding, Z. (2021). Hydrothermal synthesis and microwave absorption properties of nickel ferrite/multiwalled carbon nanotubes composites. *Coatings*, 11(5), 534.
89. Rostami, M., & Ara, M. H. M. (2019). The dielectric, magnetic and microwave absorption properties of Cu-substituted Mg-Ni spinel ferrite-MWCNT nanocomposites. *Ceramics International*, 45(6), 7606–7613.
90. Mustaffa, M. S., Azis, R. A. S., Abdullah, N. H., Ismail, I., & Ibrahim, I. R. (2019). An investigation of microstructural, magnetic and microwave absorption properties of multi-walled carbon nanotubes/Ni0. 5Zn0. 5Fe2O4. Scientific reports, 9(1), 15523.
91. Chahar, M., Dabas, S., & Thakur, O. P. (2022). Enhanced electromagnetic shielding effectiveness of MWCNT/zinc-doped nickel ferrite nanocomposites. *Ceramics International*, 48(4), 5352–5360.
92. Dalal, M., Mallick, A., Greneche, J. M., Das, D., & Chakrabarti, P. K. (2017). Correlation of cation distribution with the hyperfine and magnetic behaviour of Ni0. 3Zn0. 4Co0. 2Cu0. 1Fe2O4 nanoparticles and their microwave absorption properties when encapsulated in multi-walled carbon nanotubes. *Journal of Physics: Condensed Matter*, 29(8), 085803.
93. Gao, Y., & Wang, Z. (2021). Microwave absorption and electromagnetic interference shielding properties of Li-Zn ferrite-carbon nanotubes composite. *Journal of Magnetism and Magnetic Materials*, 528, 167808.
94. Shu, R., Zhang, G., Zhang, J., Wang, X., Wang, M., Gan, Y., ... He, J. (2018). Fabrication of reduced graphene oxide/multi-walled carbon nanotubes/zinc ferrite hybrid composites as high-performance microwave absorbers. *Journal of Alloys and Compounds*, 736, 1–11.
95. Ding, G., Chen, C., Tai, H., Tang, Z., Wang, Z., Cheng, G., & Wan, X. (2021). Structural characterization and microwave absorbing performance of CuFe2O4/RGO composites. *Journal of Solid State Chemistry*, 297, 122051.
96. Zhang, G., Shu, R., Xie, Y., Xia, H., Gan, Y., Shi, J., & He, J. (2018). Cubic MnFe2O4 particles decorated reduced graphene oxide with excellent microwave absorption properties. *Materials Letters*, 231, 209–212.
97. Yang, H., Ye, T., Lin, Y., Zhu, J., & Wang, F. (2016). Microwave absorbing properties of the ferrite composites based on graphene. *Journal of Alloys and Compounds*, 683, 567–574.

98. Sani, Y., Ismail, I., Yaakob, Y., & Mohammed, J. (2023). Enhanced electromagnetic microwave absorbing performance of carbon nanostructures for RAMs: A review. *Applied Surface Science Advances*, 18, 100455.
99. Yang, H., Ye, T., Lin, Y., Zhu, J., & Wang, F. (2016). Microwave absorbing properties of the ferrite composites based on graphene. *Journal of Alloys and Compounds*, 683, 567–574.
100. Sodaee, T., Ghasemi, A., & Razavi, R. S. (2017). Cation distribution and microwave absorptive behavior of gadolinium substituted cobalt ferrite ceramics. *Journal of Alloys and Compounds*, 706, 133–146.
101. Aman, S., Ahmad, N., Almutairi, B. S., Tahir, M. B., & Ali, H. E. (2023). Study of gadolinium substituted barium-based spinel ferrites for microwave applications. *Journal of Electronic Materials*, 52(6), 4149–4161.
102. Bhuvaneshwari, V., Lenin, N., Shiva, C., Kathirvel, M., Ragavendran, V., Rajeshwari, A., ... Kumar, G. V. (2023). Influence of gadolinium doped in nickel nanoferrites on structural, optical, electrical, and magnetic properties. *Materials Science and Engineering: B*, 288, 116184.
103. Neelima, P., Ramesh, T., Raju, P., & Murthy, S. R. (2021). Structural and microwave behavior of Dy 3+-substituted Ni 0.5 Zn 0.5 Dy x Fe 2-x O 4 ferrites. *Journal of Materials Science: Materials in Electronics*, 32, 1729–1740.
104. Almessiere, M. A., Slimani, Y., Güngüneş, H., Korkmaz, A. D., Zubar, T., Trukhanov, S., ... Baykal, A. (2021). Influence of Dy3+ ions on the microstructures and magnetic, electrical, and microwave properties of [Ni0. 4Cu0. 2Zn0. 4](Fe2–x Dy x) O4 (0.00≤ x≤ 0.04) spinel ferrites. *ACS Omega*, 6(15), 10266–10280.
105. Akhtar, M., Hasan, M. S., Amin, N., Morley, N. A., & Arshad, M. I. (2023). Tuning structural, electrical, dielectric and magnetic properties of Mg–Cu–Co ferrites via dysprosium (Dy3+) doping. *Journal of Rare Earths*, 42(1), 137–146.
106. Kumbhar, R. N., Shinde, T. J., Kamble, S. A., Mathe, V. L., & Ghodake, J. S. (2021). Influence of rare earth ions (Sm3+, Dy3+) substitution on magnetic and microwave performance of magnesium ferrite. *Physica B: Condensed Matter*, 619, 413161.
107. Akhtar, M. N., Makhdoom, S., Baqir, M. A., Yousaf, M., Khan, M. A., Batoo, K. M., & Hussain, S. (2022). Enhanced structural, electromagnetic and absorption features of CoSm ferrite-metamaterial absorbers through synergistic effects. *Ceramics International*, 48(20), 29561–29571.
108. Stergiou, C. (2017). Magnetic, dielectric and microwave absorption properties of rare earth doped Ni–Co and Ni–Co–Zn spinel ferrites. *Journal of Magnetism and Magnetic Materials*, 426, 629–635.
109. Srinivas, C., Praveen, K. N., Kumar, E. R., Singh, S., Meena, S. S., Bhatt, P., ... Sastry, D. L. (2022). Microwave absorption properties of rare earth (RE) ions doped Mn–Ni–Zn nanoferrites (RE= Dy, Sm, Ce, Er) to shield electromagnetic interference (EMI) in X-band frequency. *Ceramics International*, 48(22), 33891–33900.
110. Mandal, D., & Mandal, K. (2021). Electromagnetic wave attenuation properties of MFe2O4 (M= Mn, Fe, Co, Ni, Cu, Zn) nano-hollow spheres in search of an efficient microwave absorber. *Journal of Magnetism and Magnetic Materials*, 536, 168127.
111. Tong, Z., Yao, Q., Deng, J., Cheng, L., Chuang, T., Wang, J., ... Wang, Z. (2021). Effects of Ni-doping on microstructure, magnetic and microwave absorption properties of CoFe2O4. *Materials Science and Engineering: B*, 268, 115092.
112. Kaur, M., Kaur, P., & Bahel, S. (2022). K-band microwave absorption analysis of sol–gel synthesized cobalt-substituted zinc spinel ferrites. *Journal of Materials Science: Materials in Electronics*, 33(15), 12182–12200.

113. Ünal, B., Ünver, İ. S., Güngüneş, H., Topal, U., Baykal, A., & Sözeri, H. (2016). Microwave, dielectric and magnetic properties of Mg-Ti substituted Ni-Zn ferrite nanoparticles. *Ceramics International*, 42(15), 17317–17331.
114. Aggarwal, N., & Narang, S. B. (2021). Effect of co-substitution of Co–Zr on electromagnetic properties of Ni–Zn spinel ferrites at microwave frequencies. *Journal of Alloys and Compounds*, 866, 157461.
115. Gao, Y., Wang, Z., Shi, R., Pei, J., Zhang, H., & Zhou, X. (2019). Electromagnetic and microwave absorption properties of Ti doped Li–Zn ferrites. *Journal of Alloys and Compounds*, 805, 934–941.
116. Teber, A., Cil, K., Yilmaz, T., Eraslan, B., Uysal, D., Surucu, G., ... Bansal, R. (2017). Manganese and zinc spinel ferrites blended with multi-walled carbon nanotubes as microwave absorbing materials. *Aerospace*, 4(1), 2.
117. Bao, X., Wang, X., Zhou, X., Shi, G., Xu, G., Yu, J., ... Choi, C. (2018). Excellent microwave absorption of FeCo/ZnO composites with defects in ZnO for regulating the impedance matching. *Journal of Alloys and Compounds*, 769, 512–520.
118. Shu, R., Zhang, G., Wang, X., Gao, X., Wang, M., Gan, Y., ... He, J. (2018). Fabrication of 3D net-like MWCNTs/ZnFe2O4 hybrid composites as high-performance electromagnetic wave absorbers. *Chemical Engineering Journal*, 337, 242–255.
119. Zhang, J., Shu, R., Guo, C., Sun, R., Chen, Y., & Yuan, J. (2019). Fabrication of nickel ferrite microspheres decorated multi-walled carbon nanotubes hybrid composites with enhanced electromagnetic wave absorption properties. *Journal of Alloys and Compounds*, 784, 422–430.
120. Golchinvafa, S., Masoudpanah, S. M., & Jazirehpour, M. (2019). Magnetic and microwave absorption properties of FeCo/CoFe2O4 composite powders. *Journal of Alloys and Compounds*, 809, 151746.
121. Aggarwal, N., & Narang, S. B. (2021). Microwave absorption analysis of Mg–Zr-substituted Ni–Zn spinel ferrites in the X-band. *Journal of Physics and Chemistry of Solids*, 159, 110289.
122. Yan, Z., & Luo, J. (2017). Effects of CeZn co-substitution on structure, magnetic and microwave absorption properties of nickel ferrite nanoparticles. *Journal of Alloys and Compounds*, 695, 1185–1195.
123. Lei, C., & Du, Y. (2020). Tunable dielectric loss to enhance microwave absorption properties of flakey FeSiAl/ferrite composites. *Journal of Alloys and Compounds*, 822, 153674.
124. Aggarwal, N., & Narang, S. B. (2022). Synthesis of Ni-Zn-Mg-Zr spinel ferrites as high-performance electromagnetic wave absorber in Ku band. *Materials Today: Proceedings*, 52, 1294–1301.
125. Aggarwal, N., & Narang, S. B. (2021). Microwave absorption analysis of Mg–Zr-substituted Ni–Zn spinel ferrites in the X-band. *Journal of Physics and Chemistry of Solids*, 159, 110289.
126. Kaur, M., & Bahel, S. (2023). Microwave absorption characteristics of cobalt doped zinc spinel ferrites in Ku-band (12.4–18 GHz). *Materials Today: Proceedings*, 82, 1–7.
127. Gholipur, R. (2022). Electrical, magnetic and absorption study in the Mn1-xZrxFe2O4 nanocrystallites. *Journal of Materials Science: Materials in Electronics*, 33(16), 13180–13189.

4 Magnetic Properties of Spinel and Inverse Spinel Ferrites

Deepika Sharma and Kanchan Bala

4.1 INTRODUCTION

Numerous research studies have focused on ferrite-based nanoparticles, particularly magnetic nanoparticles exhibiting spinel and inverse ferrites. These materials hold significant importance in fundamental sciences, elucidating the intricate relationships between magnetic properties and their crystal structures, and the chemistry of nanoparticles. Ferrites are classified into various groups based on their magnetic properties and crystal structure, including orthoferrites, hexaferrites, garnets, and spinel ferrites. Spinel ferrite, characterized by magnetic material having the chemical formula AB_2O_4, involves different metal ions residing in tetrahedral and octahedral sites denoted by A and B, respectively [1]. Spinel ferrites, which are characterized by a face-centered cubic structure, demonstrate ferrimagnetic behavior. Within the crystal lattice of spinel ferrite nanoparticles, three recognized structures exist based on the arrangement of +2 and +3 cations: normal, mixed, and inverse spinel ferrites. Inverse ferrites, categorized as garnet ferrites, adopt a crystal structure derived from the garnet structure. Inverse ferrite is a class with the chemical formula $A_2B_5O_{12}$, in which A and B denote distinct metal ions occupying dodecahedral and octahedral sites, respectively [2]. Inverse ferrites find applications in magneto-optical devices, magnetic sensors, and microwave devices. They boast distinctive magnetic properties such as high-frequency operational capacity and low anisotropy.

4.2 MAGNETIC PROPERTIES OF FERRITES

Magnetism in ferrites results from the irregularly distributed 3D electrons in disproportionate numbers at octahedral and tetrahedral sites. Based on Néel's two sublattice model, spinel ferrites exhibit a ferrimagnetic type (i.e., anti-ferromagnetic type) of order, in which the magnitude is not equal, but the magnetic moments are opposite to each other [3, 4]. Therefore, these types of materials exhibit a non-zero net magnetization. Yafet and Kittel extended this theory by explaining that the octahedral and tetrahedral sites are partitioned into sublattices [5]. Vector resultants of

 DOI: 10.1201/9781032665658-4

the magnetic moments in these sublattices are associated opposite to each other, leading to ferrimagnetism.

Spinel ferrites possess outstanding physical characteristics, such as large surface areas, significant chemical stability, and enhanced magnetic properties. The magnetic assets of spinel ferrites are tailored by adjusting composition and synthesis conditions. Spinel ferrites have superparamagnetic assets at the nanoparticle scale, along with high chemical and thermal stability, simple chemical configuration, and extensive implementation across numerous fields. Spinel ferrites are used in various devices including inductors, microwave devices, data storage devices, sensing materials, ferrofluid and catalysis, anode materials of lithium-ion batteries, and magnetic recording media [6]. The cation distribution affects these properties by doping, such as substituting one or more ions from a spinel lattice with other metal ions. At the nanoscale level, each magnetic nanoparticle behaves like an enormous paramagnetic atom, exhibiting a swift response to the fundamental magnetic fields and unique magnetic assets. The most significant magnetic assets of nanoparticles are coercivity (H_c), remanence magnetization (M_r), saturation magnetization (M_s), squareness ratio (SQ), and magnetic anisotropy [7]. The net magnetization is given by the following formula:

$$M_{\text{total}} = M_{\text{B-site}} - M_{\text{A-site}} \tag{4.1}$$

where, $M_{\text{B-site}}$ and $M_{\text{A-site}}$ represent the net magnetization of octahedral and tetrahedral sites, respectively.

4.2.1 Saturation Magnetization

Saturation magnetization is a crucial property of spinel ferrites, influenced by the cationic dispersal between tetrahedral and octahedral sites. The combination of different metal ions results in different values of saturated magnetization, which are determined by inserting the M–H curve as H approaches ∞ [8]. The optimal magnetic moment of magnetic nanoparticles in an external magnetic field, resulting from the parallel alignment of unpaired electrons, is known as M_s. The theoretical value of the saturation magnetization for a single-phase ferrite structure may be calculated by the formula [9]:

$$\mathrm{M}_s = \frac{8n\mu_B}{10^3 a^3 \rho} \tag{4.2}$$

where

n = theoretical number of magnetic moments per formula unit
μ_B = magnitude of Bohr magneton
a = lattice parameter
ρ = density

Spinel ferrites possess higher M_s values, which makes them useful. Prasad et al. developed nanocrystalline ferrites MFe_2O_4, where M = Co, Ni, Cu, and Zn [10]. Among these spinel ferrites, $CoFe_2O_4$ exhibited the highest saturation magnetization having a value of 41 emu/g, on the other hand, $ZnFe_2O_4$ had the lowest saturation magnetization having a value of 12.87 emu/g. Fe-based spinel ferrites (Fe_3O_4) have higher M_s as compared to Ni- or Mn-based ferrites ($NiFe_2O_4$, $MnFe_2O_4$) [11]. The unit cell structure of magnetic spinel and inverse spinel ferrites is shown in Figure 4.1.

Mirzaee et al. reported manganese-doped Ni-ferrite nanoparticles, in which magnetic dimensions revealed that as Mn ion concentration increased, the M_s value of the samples decreased due to Mn ion dispersion in the spinel sites and structure strain [12, 13]. Different factors affect the value of M_s of doped ferrites (Figure 4.2).

1. **Magnetic nature:** The magnetic nature of the doping material affects the value of M_s, for example, ions having a greater magnetic moment (Fe^{3+} with a magnetic moment of 5μB) are replaced by ions having a lesser magnetic moment (Mn^{2+}, Ti^{4+} with magnetic moment 0 μB). This results in a reduction in the saturation magnetization behavior and *vice versa*. The rise in saturation magnetization observed when doping of Co in Ni-ferrites is attributed to changes in the exchange interaction among the octahedral and tetrahedral sublattices, enhanced crystallinity, and small particle dimension distribution.
2. **Crystallite and particle size:** The M_s is affected by the crystallite size and shape of ferrite particles. A small proportion of crystallite particles tends to have a great surface-to-volume ratio and leads to enhanced magnetic properties. Zhang et al. suggested that the M_s value of $CoFe_2O_4$ increases with an increase in crystallite extent that is held by the dead layer, consistent with the core–shell model theory [14]. It was suggested that when the crystallite size becomes negligible, the magnetic nanoparticles are segregated by the nonmagnetic layer, while the dead layers represent surface spin disarray. Therefore, the rise in surface spin observed with a reduction

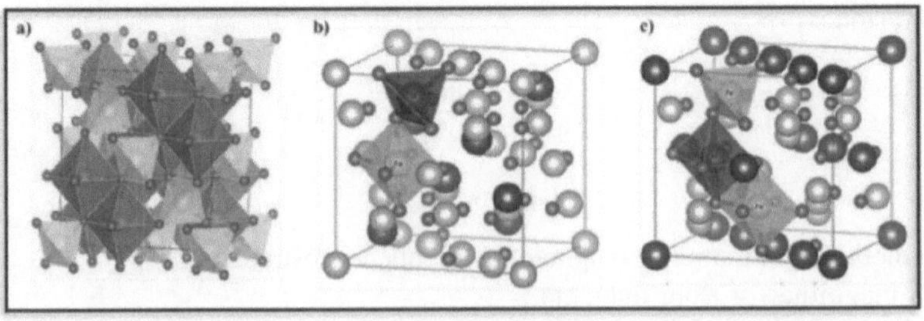

FIGURE 4.1 Unit cell structure of magnetic spinel ferrites: (a) tetrahedral sites (yellow), octahedral sites (blue), oxygen atom (red); (b) normal spinel ferrite; and (c) inverse spinel ferrites [11].

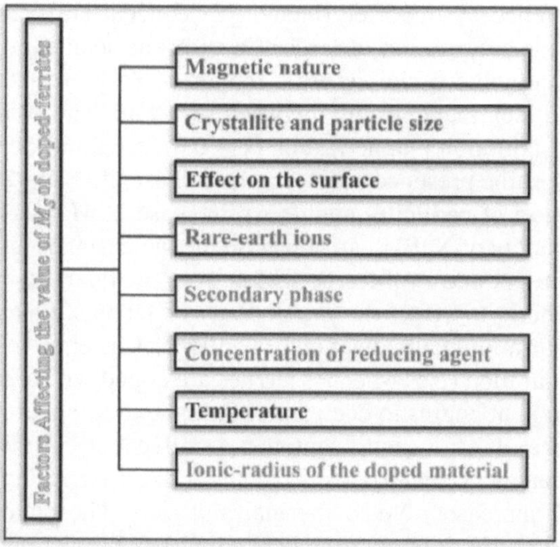

FIGURE 4.2 Different factors affecting the value of M_s of doped ferrites.

in crystallinity plays a major role in the lowering of the saturation magnetization. Sivakumar et al. have studied that the saturation magnetization of a sample with 72 nm grain size has improved by 33% compared to that of micron-sized particles [15]. Makovec et al. have also described in their research that the high saturation magnetization is attributed to good crystallinity of the nanoparticles [16].

3. **Effect on the surface:** The reduction in saturation migration has been attributed to the occurrence of magnetically depleted layers on the material's surface, and the spin-slanting effect throughout the whole volume of the crystallite [17].
4. **Rare-earth ions:** The incorporation of rare-earth ions results in magnetic dilution and spin–orbit coupling, causing a shift to a collinear ferrimagnetic spin configuration on B-sites [18]. When rare-earth ions occupy octahedral sites, they distort the collinear spin arrangement, which results in a reduction in M_s. The doping of spinel lattices with rare-earth cations may enhance the nano-magnetism phenomenon [19]. Additionally, the interaction between rare-earth metal and spinel ferrites (Re^{3+}-Fe^{3+}) helps to fine-tune the magnetic assets [20]. Ni-Zn nanoferrites doped with rare-earth metal cations are expected to exhibit superior mechanical, electrical, physical, and chemical properties. These nanoparticles could also be appropriate for use in memory devices, information storage, high-frequency electronic devices, and microwave devices [21].
5. **Secondary phase:** The doped ferrite's M_s value is influenced by the occurrence of the secondary phase. This secondary phase hinders the growth of the spinel lattice and the expansion of grains in spinel ferrites. The secondary

phase occurrence was reported when the Ce concentration exceeds 0.6 [22], because of the substitution of larger Ce^{3+} ions at octahedral sites. Ni ions typically occupy the B-site, whereas the ion of Zn prefers site A, and ions of Fe^{3+} are disseminated evenly between site A and site B, respectively. A decrease in saturation magnetization ($x \geq 0.3$) in indium-doped Ni-ferrites is attributed to the presence of a secondary phase of In_2O_3 [23].

6. **Concentration of reducing agent:** An increase in M_s was observed with the rising amount of $NaBH_4$, attributed to the increased occupancy of metal cations in sites A and B of ferrites. Moreover, the quantity of the reducing agent was found to affect the morphology of ferrites, potentially leading to higher saturation magnetization [24]. Vigneswari et al. investigated that magnetic saturation rises when Ni-ferrites are doped with a 20% concentration of Ca ion but begins to decrease with higher doping levels [25].
7. **Temperature:** It is a significant factor influencing the M_s of ferrites. Generally, magnetic materials experience a decrease in magnetization as temperature increases due to thermal agitation. The doped ferrites can have altered Curie temperatures depending on the doping elements and concentration.
8. **Ionic radius of the doped material:** The ionic radius of the doped material significantly influences determining saturation magnetization in ferrites [20]. When the ionic radius of a dopant with lower magnetic properties or nonmagnetic character exceeds that of the host or substituted cations, the AB interaction weakens, which results in a reduction in M_s.

4.2.2 Coercivity

Coercivity refers to the resistance of ferromagnetic materials to demagnetization. This property is used to classify magnets as soft or hard. Materials with high coercivity are considered hard magnetic materials because they require a strong magnetic field to demagnetize them (e.g., neodymium-iron-boron or samarium-cobalt). However, the materials with low coercivity are soft magnetic materials. These materials can magnetize easily and demagnetize with weak magnetic fields. In the case of water purification, soft magnets are notable for their fine magnetic hysteresis loops and relatively low H_c value, whereas hard magnets are characterized by a wide hysteresis loop and high coercivity value [26]. Coercivity measures the resistance to demagnetization of the magnetic materials. Simply, it is a measure to know how difficult it is to erase or change the magnetic state of a material.

Coercive forces vary inversely with grain dimensions; as the grain dimensions diminish, the volume concentration of grain margins increases [27]. This, in turn, boosts the holding capacity of magnetic domain walls, leading to enhanced reversibility in magnetization. The occurrence of a secondary phase around the grain margins in spinel ferrites impedes domain-wall displacement, resulting in increased coercivity [28]. Meng et al. reported in their study that the magnetization process can be influenced by the demagnetizing field, consequently causing a decrease in coercivity [18]. This is attributed to a higher presence of magnetic domains and domain walls, facilitating easier demagnetization. Additionally, a rise in porosity imposes a stronger field to push the domain wall, resulting in increased coercivity. Generally, coercivity depends on grain size, particle size, sintering temperature, microstrain, magnetocrystalline anisotropy, magnetic particle morphology, and saturation magnetization, which can be stated as

$$\mathrm{H}_C = \frac{2K_1}{\mu_0 M_s} \tag{4.3}$$

$$\mathrm{K1} = \mathrm{H}_C \times \frac{M_S}{0.96} \tag{4.4}$$

where μ_0 and M_s are the free space permeability and saturation magnetization respectively [29].

As the grain size decreases with decreasing sintering temperature, the coercivity will increase due to dependence on grain size [30]. Soda et al. justified that the reduction of particle size contributes to an increase in coercivity and a decrease during sintering [31]. Kannon et al. reported in their study that the coercivity decreases with rising sintering temperature because a large grain size emerges, which leads to more internal walls within each grain [32]. Consequently, the influence on magnetization or demagnetization intensifies due to increased wall motion. Kaur et al developed Cr-substituted cobalt nanoferrites with composition [33]. As chromium content increases, coercivity steadily declines and is accompanied by some cobalt ions transitioning from octahedral sites to tetrahedral sites. Hence, it was reported that the coercivity decrease is linked to increased chromium concentration. Coercivity values increase in direct proportion to the magnetocrystalline anisotropy constant (K). The reduced coercivity caused by heightened Cr content shifts the material from a hard to a soft magnetic substance.

4.2.3 Remanence Magnetization

Remanence magnetization is also known as remnant or residual magnetization. It refers to the magnetic field that remains in the material after the external magnetic field has been withdrawn. It represents the permanent magnetization of the material, which leads to the alignment of magnetic domains within it. Materials with low or no remnant magnetization are referred to as nonpermanent or temporary magnets. Remanent magnetization is utilized in innumerable applications such as hard drives, magnetic tapes, magnetic separators, magnetic sensors, electric motors, and generator construction. The magnitude of remnant magnetization is influenced by abundant factors like the structure, composition, and coercivity of the materials. Materials with high coercivity retain strong remnant magnetization.

Spinel ferrites possessing negligible remanence M_r, high M_s, and low H_c are well-suited for water purification purposes [2]. Figure 4.3 illustrates a schematic diagram depicting essential magnetic properties like M_r, M_s, and H_c. Nanoparticles exhibiting a superparamagnetic state play a crucial role in the treatment of water due to

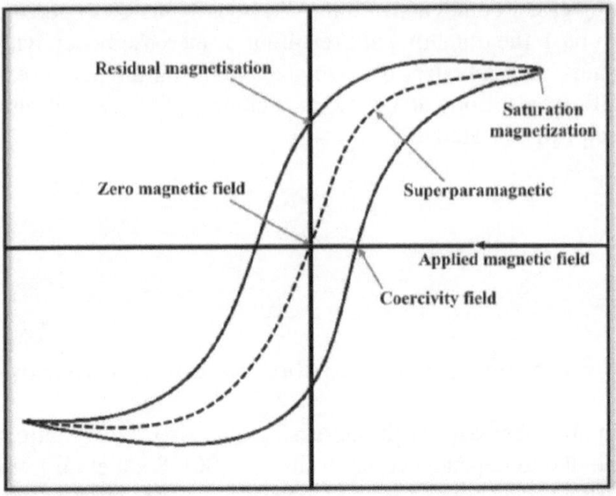

FIGURE 4.3 Schematic diagram of magnetic properties of magnetic nanoparticles.

their ability to operate without the need for an external magnetic field [34]. These nanoparticles neither clump together nor retain residual magnetism, allowing them to disperse freely in water. However, upon applying the magnetic field, they transform into a superparamagnetic state, facilitating easy separation. The phenomenon of superparamagnetism typically manifests in small ferromagnetic nanoparticles ranging in size from 3 to 50 nm. Single-domain nanoparticles can undergo random directional changes in response to temperature fluctuations [35].

Squareness ratio:

The value of SQ reflects the magneto-crystalline anisotropy and superexchange interactions inherent to the studied nanoferrites and is given by the following formula:

$$SQ = M_r/M_s \tag{4.2}$$

The squareness ratio is highly influenced by factors such as magnetic anisotropy, crystal defects, particle proportions, density, and synthesis techniques in the investigated nano ferrites. An SQ nearing 0.5 suggests the presence of pseudo-single-domain particles within the materials. An SQ value below 0.5 indicates a composition of multi-domain-type ferrites [18]. An SQ value of less than 0.1 was reported by Khan, supporting the presence of multi-domain grain [36]. Junaid et al. reported the SQ ratio of all the synthesized nanocrystalline ferrites had an SQ ratio below 0.2, indicating that substituted Cu-ferrites constitute single-domain ferrites material [37].

4.2.4 Magnetic Anisotropy

Another significant term utilized to elucidate the magnetic properties of nanoparticles is magnetic anisotropy. This property refers to a magnetic material exhibiting preferential alignment of its magnetic moments along a specific crystallographic direction or in a particular orientation. The spin–orbit interaction is the primary source of magnetic anisotropy, particularly in isolated systems, as it governs the direction of spin alignment. Magnetic anisotropy directly influences key magnetic assets or properties like M_s, M_r and H_c. Puli et al. reported polycrystalline gadolinium-substituted cobalt ferrite ceramics synthesized via a solid-state reaction technique. They attributed the increase in coercivity with the higher Gd^{3+} content to the heightened magnetic anisotropy within the ceramic [38]. Muscas et al. explored the adaptability of temperature thermal decomposition synthesis in the production of magnetic nanoparticles (Fe_3O_4 and $CoFe_2O_4$) with precise control over particle dimensions within the range of 5 to 8 nm [39]. This method enables adjustment not only of particle dimensions but also of their magnetic anisotropy for specific application purposes. The replacement of Fe with Co resulted in a significant growth in magnetocrystalline anisotropy. They demonstrated that magnetocrystalline anisotropy predominantly influences the magnetic behavior of the particles, except for 5 nm Fe_3O_4, where surface disorder contributes significantly to the anisotropy.

4.3 MAGNETIC PROPERTIES OF SOME IMPORTANT FERRITE NANOPARTICLES

Nanostructure spinel ferrites exhibit intriguing electrical and magnetic properties, alongside a high surface-to-volume ratio, that make them increasingly appealing for applications in catalysis, sensors, and biomedicines [40]. The reduced dimensions of magnetic nanoparticles contribute to an increase in coercivity. There is considerable interest in devising cost-effective and high-yield processes for manufacturing single-phase spinel ferrite magnetic nanoparticles with manageable dispersal of grain proportions.

4.3.1 Nickel Ferrite Nanoparticles

Ni-ferrites have various interesting properties such as low coercivity, moderate saturation magnetization, high electrical resistivity, excellent chemical stability, and low electrical losses. These attributes make them appropriate for various applications in environmental, technological, and biomedical fields. Ni-ferrites possess an inverse spinel structure, with $8Fe^{3+}$ ions occupying the tetrahedral A-sites and other $8Fe^{3+}$ ions composed with $8Ni^{2+}$ ions located in octahedral B-sites within the unit cell. Anumol et al. have synthesized magnesium-substituted Ni-ferrite nanoparticles ($Mg_xNi_{1-x}Fe_2O_4$; $0 \leq x \leq 1$), using sol–gel technique, followed by annealing at 550°C and 900°C in air [41]. It was observed that the crystallite size remained unchanged with increasing Mg^{2+} concentration but increased after annealing in the nanoparticle samples. The maximum saturation magnetization values of 42 and 46 emu/g were recorded at 300 and 60 K, respectively, for the sample annealed at 550°C with $x = 1$. Additionally, coercivity and magnetization diminished with a higher concentration of Mg^{2+}.

Manikandan et al. reported the ruthenium-doped Ni-ferrite ($NiFe_2O_4$) nanoparticles in their research [42]. In magnetic studies, they revealed that the arrangement of Ni, Fe, and Ru ions at the octahedral and tetrahedral sites changes the magnetization dynamics in spinel ferrites. It was observed that the magnetization increases with higher annealing temperature. Saturation magnetization and coercivity were observed to lower because of the favorable effect of Ru doping in $NiFe_2O_4$ on its soft magnetic behavior. The magnetic moments of Ni, Fe, and Ru ions 2, 5, and 4 μB, respectively, demonstrate that Ni^{2+} is situated in A-sites, while Ru^{3+} and Fe^{3+} are positioned in the B-sites, leading to an increase in overall magnetization.

4.3.2 Magnesium Ferrite Nanoparticles

Mg-ferrites ($MgFe_2O_4$) possess an inverse spinel structure, in which Mg^{2+} and Fe^{3+} cations are dispersed unevenly between A- and B-sites. Mg-ferrites have some important and interesting applications, including in heterogeneous catalysis, biomedicines, and magnetic technologies [43, 44]. These soft magnetic materials are easily magnetized as well as demagnetized. Sivakumar et al. reported nanocrystalline $MgFe_2O_4$ with grain sizes ranging from 19nm to 72 nm, using a ceramic technique [15]. Observations demonstrated that sample with 72 nm grain dimensions exhibited the maximum saturation magnetization (36 emu/g) at 293 K temperature. $Mg_xZn_{1-x}Fe_2O_4$ (where $x = 0.7, 0.5, 0.25$) nanoparticles were prepared using three different methods: auto-combustion, coprecipitation, and spray pyrolysis [45]. The maximum specific magnetization was observed in $Mg_{0.5}Zn_{1-0.5}Fe_2O_4$ nanoparticles at room temperature, obtained by the citrate auto-combustion method (30 emu/g). Kuru et al. investigated the structural, impedance, dielectric, and humidity properties of Cr^{3+}-substituted Mg-Zn ferrites using the chemical co-precipitation method [46]. The study found that with the increase in Cr^{3+} substitution, the saturation magnetization decreased while coercivity increased. The Mg-Zn-Cr ferrites show relaxor ferroelectric properties, with their Curie temperature shifting lower with Cr^{3+} addition, and their impedance varying with temperature and humidity.

4.3.3 Zinc Ferrite Nanoparticles

Zn-ferrites have attracted considerable attention due to their unique properties. They exhibit moderate saturation magnetization, high electromagnetic performance, and unique chemical stability. These characteristics make them highly adaptable as soft ferrites. One of the key advantages of Zn-ferrites is their high resistance and low eddy current loss. This makes them particularly useful in applications that require high efficiency and low power loss [47]. Furthermore, their ability to function as high-quality filters in radio frequency circuits adds to their versatility [48]. It has been investigated that doping can significantly alter the properties of nanoparticles [49]. For instance, doping Zn-ferrite nanoparticles with PEG has been found to improve their magnetic properties, making them suitable for applications like hyperthermia and drug delivery. The study found that crystalline size also played a role in increasing the M_s value as the effect of thermal agitation in the molecules.

Ullah Rather et al. investigated aluminium-doped Zn-ferrite nanoparticles ($ZnFe_{2-x}Al_xO_4$) with a 0–0.9 range of x via a thermal treatment method using polyvinylpyrrolidone [47]. The study found that aluminum doping influenced the optical, structural, and magnetic assets of Zn-ferrites. The magnetic assets of Zn-ferrite nanoparticles were investigated using a vibrating sample magnetometer. The results showed that samples of pure Zn-ferrite, and aluminum-doped Zn-ferrite exhibited paramagnetism. It was reported that the saturation magnetization decreased while coercivity increased with an increase in the concentration of Al. The saturation magnetization decreases with an increase in the concentration of Al. It leads to the assimilation of Zn ions, which are nonmagnetic in the octahedral sublattice, resulting in a reduction of magnetization.

4.3.4 Cobalt Ferrites Nanoparticles

Cobalt ferrite nanoparticles have garnered significant interest due to their diverse applications, including magnetic recording, paramagnetic magnets, and medical fields like molecular imaging and magnetic hyperthermia. Ghorbani et al. synthesized Co-ferrite nanoparticles doped with cadmium using a hydrothermal method [50]. The particle dimensions of the synthesized nanoparticles were found to be in the range of 56 nm to 59 nm. The lattice parameter was observed to increase from 0 to 0.5 × value. In terms of magnetic properties, the saturation magnetization initially increased to a level of 0.1. However, it was noted to decrease with an increase in the amount of Cd doping. The coercivity value was measured to be 1.1 KOe for $x = 0$, but this value was found to decrease with Cd doping. A higher concentration of Co^{2+} on the octahedral site led to an advanced saturation magnetization value of 78 emu/g, as found in a study [51]. This was detected at the medium milling energy level. Furthermore, the maximum coercivity values, which reached 2090 Oe, were obtained for the sample milled at a high-level energy. These findings contribute to the understanding of how mechanical milling energy levels can influence the magnetic properties of cobalt ferrite nanoparticles.

4.4 CONCLUSION

This chapter highlights that ferrite-based nanoparticles, especially those exhibiting spinel and inverse ferrite structures, are critical in advancing our understanding of the relationship between magnetic properties, crystal structure, and nanoparticle chemistry. The magnetic properties of ferrites, including coercivity, remanence magnetization, and saturation magnetization, are crucial for various applications. Factors such as crystalline size, surface effects, doping with rare-earth ions, secondary phases, reducing agent concentration, and temperature significantly influence these properties. Spinel ferrites exhibit superparamagnetic behavior at the nanoscale, making them suitable for applications like water purification, data storage, and medical fields. The magnetic properties and applications of specific ferrite nanoparticles like nickel, magnesium, zinc, and cobalt ferrites highlight their versatility. Ni-ferrites are noted for their low coercivity and moderate saturation magnetization, making them suitable for environmental and biomedical applications. Mg-ferrites, with their soft magnetic properties, are used in catalysis and medical fields. Zn-ferrites exhibit high resistance and low eddy current loss, making them ideal for radio frequency circuits. Co-ferrites, with their high saturation magnetization and coercivity, are useful in magnetic recording and medical applications.

REFERENCES

1. Mathew, D. S., & Juang, R. S. (2007). An overview of the structure and magnetism of spinel ferrite nanoparticles and their synthesis in microemulsions. *Chemical Engineering Journal, 129*(1–3), 51–65.
2. Vinosha, P. A., Manikandan, A., Preetha, A. C., Dinesh, A., Slimani, Y., Almessiere, M. A., ... Nirmala, G. F. (2021). Review on recent advances in synthesis, magnetic properties, and water treatment applications of cobalt ferrite nanoparticles and nanocomposites. *Journal of Superconductivity and Novel Magnetism, 34*, 995–1018.
3. Néel, L. (1984). Magnetic properties of ferrites: Ferrimagnetism and antiferromagnetism. *Physical Chemical & Earth Sciences, 31*, 18.
4. Ünal, B., Ünver, İ. S., Güngüneş, H., Topal, U., Baykal, A., & Sözeri, H. (2016). Microwave, dielectric, and magnetic properties of Mg-Ti substituted Ni-Zn ferrite nanoparticles. *Ceramics International, 42*(15), 17317–17331.
5. Yafet, Y., & Kittel, C. (1952). Antiferromagnetic arrangements in ferrites. *Physical Review, 87*(2), 290.
6. Manohar, A., Krishnamoorthi, C., Naidu, K. C. B., & Pavithra, C. (2019). Dielectric, magnetic hyperthermia, and photocatalytic properties of ZnFe 2 O 4 nanoparticles synthesized by solvothermal reflux method. *Applied Physics A, 125*, 1–10.
7. Wahba, A. M., & Mohamed, M. B. (2015). Structural and magnetic characterization and cation distribution of nanocrystalline CoxFe3– xO4 ferrites. *Journal of Magnetism and Magnetic Materials, 378*, 246–252.
8. Kumar, L., & Kar, M. (2011). Influence of Al3+ ion concentration on the crystal structure and magnetic anisotropy of nanocrystalline spinel cobalt ferrite. *Journal of Magnetism and Magnetic Materials, 323*(15), 2042–2048.
9. Valenzuela, R. (2012). Novel applications of ferrites. *Physics Research International 2012*(1), 591839.
10. Vara Prasad, B. B. V. S., Ramesh, K. V., & Srinivas, A. (2017). Structural and magnetic studies of nano-crystalline ferrites MFe_2O_4 (M = Zn, Ni, Cu, and Co) synthesized via citrate gel autocombustion method. *Journal of Superconductivity and Novel Magnetism, 30*, 3523–3535.
11. Salih, S. J., & Mahmood, W. M. (2023). Review on magnetic spinel ferrite (MFe_2O_4) nanoparticles: From synthesis to application. *Heliyon, 9*(6), e16601.
12. Mirzaee, S., Azizian-Kalandaragh, Y., & Rahimzadeh, P. (2020). Modified co-precipitation process effects on the structural and magnetic properties of Mn-doped nickel ferrite nanoparticles. *Solid State Sciences, 99*, 106052.
13. Rana, G., Dhiman, P., Kumar, A., Vo, D. V. N., Sharma, G., Sharma, S., & Naushad, M. (2021). Recent advances on nickel nano-ferrite: A review on processing techniques, properties, and diverse applications. *Chemical Engineering Research and Design, 175*, 182–208.
14. Zhang, Y., Yang, Z., Yin, D., Liu, Y., Fei, C., Xiong, R., ... Yan, G. (2010). Composition and magnetic properties of cobalt ferrite nanoparticles prepared by the co-precipitation method. *Journal of Magnetism and Magnetic Materials, 322*(21), 3470–3475.

15. Sivakumar, N., Narayanasamy, A., Greneche, J. M., Murugaraj, R., & Lee, Y. S. (2010). Electrical and magnetic behavior of nanostructured MgFe2O4 spinel ferrite. *Journal of Alloys and Compounds, 504*(2), 395–402.
16. Makovec, D., Košak, A., Žnidaršič, A., & Drofenik, M. (2005). The synthesis of spinel–ferrite nanoparticles using precipitation in microemulsions for ferrofluid applications. *Journal of Magnetism and Magnetic Materials, 289*, 32–35.
17. Ren, X., & Xu, G. (2014). Electromagnetic and microwave absorbing properties of NiCoZn-ferrites doped with La3+. *Journal of Magnetism and Magnetic Materials, 354*, 44–48.
18. Ditta, A., Khan, M. A., Junaid, M., Khalil, R. A., & Warsi, M. F. (2017). Structural, magnetic, and spectral properties of Gd and Dy co-doped dielectrically modified Co-Ni (Ni0. 4Co0. 6Fe2O4) ferrites. *Physica B: Condensed Matter, 507*, 27–34.
19. Akhtar, M. N., & Khan, M. A. (2018). Effect of rare earth doping on the structural and magnetic features of nanocrystalline spinel ferrites prepared via sol-gel route. *Journal of Magnetism and Magnetic Materials, 460*, 268–277.
20. Akhtar, M. N., Ali, K., Umer, A., Ahmad, T., & Khan, M. A. (2018). Structural elucidation, and morphological and magnetic behavior evaluations, of low-temperature sintered, Ce-doped, nanostructured garnet ferrites. *Materials Research Bulletin, 101*, 48–55.
21. Ishaque, M., Islam, M. U., Khan, M. A., Rahman, I. Z., Genson, A., & Hampshire, S. (2010). Structural, electrical, and dielectric properties of yttrium substituted nickel ferrites. *Physica B: Condensed Matter, 405*(6), 1532–1540.
22. Krupa, J. C., & Queffelec, M. (1997). UV and VUV optical excitations in wide band gap materials doped with rare earth ions: 4f–5d transitions. *Journal of Alloys and Compounds, 250*(1–2), 287–292.
23. Meng, F., Shang, X., Zhang, W., & Zhang, Y. (2017). Effect of In3+ substitution on structural and magnetic properties of Ni ferrite nanoparticles. *Physica B: Condensed Matter, 504*, 69–73.
24. Saravanakumar, B., Rani, B. J., Ravi, G., Thambidurai, M., & Yuvakkumar, R. (2017). Reducing agent (NaBH4) dependent structure, morphology, and magnetic properties of nickel ferrite ($NiFe_2O_4$) nanorods. *Journal of Magnetism and Magnetic Materials, 428*, 78–85.
25. Vigneswari, T., & Raji, P. (2017). Structural and magnetic properties of calcium doped nickel ferrite nanoparticles by co-precipitation method. *Journal of Molecular Structure, 1127*, 515–521.
26. Murtaza, A., Yang, S., Khan, M. T., Ghani, A., Zhou, C., & Song, X. (2018). Temperature-dependent magnetization and coercivity in morphotropic phase boundary involved ferromagnetic $Tb_{1-x}Gd_xFe_2$ system. *Materials Chemistry and Physics, 217*, 278–284.
27. Van der Zaag, P. J., Van der Valk, P. J., & Rekveldt, M. T. (1996). A domain size effect in the magnetic hysteresis of NiZn-ferrites. *Applied Physics Letters, 69*(19), 2927–2929.
28. Sarac, M. F. (2020). Magnetic, structural, and optical properties of gadolinium-substituted $Co_{0.5}Ni_{0.5}Fe_2O_4$ spinel ferrite nanostructures. *Journal of Superconductivity and Novel Magnetism, 33*(2), 397–406.
29. Humbe, A. V., Kharat, P. B., Nawle, A. C., & Jadhav, K. M. (2018). Nanocrystalline $Ni_{0.70}-_xCu_xZn_{0.30}Fe_2O_4$ with $0 \leq x \leq 0.25$ prepared by nitrate-citrate route: Structure, morphology and electrical investigations. *Journal of Materials Science: Materials in Electronics, 29*, 3467–3481.
30. Stergiou, C. (2016). Microstructure and electromagnetic properties of Ni-Zn-Co ferrite up to 20 GHz. *Advances in Materials Science and Engineering, 2016*(1), 1934783.

31. Soda, R., Takagi, K., Jinno, M., Yamaguchi, W., & Ozaki, K. (2016). Anisotropic Sm2Fel7N3 sintered magnets without coercivity deterioration. *AIP Advances*, *6*(11), 115108.
32. Kannan, Y. B., Saravanan, R., Srinivasan, N., & Ismail, I. (2017). Sintering effect on structural, magnetic, and optical properties of $Ni_{0.5}Zn_{0.5}Fe_2O_4$ ferrite nano particles. *Journal of Magnetism and Magnetic Materials*, *423*, 217–225.
33. Kaur, S., Chalotra, V. K., Jasrotia, R., Bhasin, V., Kumari, S., Thakur, S., ... Kumar Godara, S. (2022). Spinel nano ferrite ($CoFe_2O_4$): The impact of Cr doping on its structural, surface morphology, magnetic, and antibacterial activity traits. *Optical Materials*, *133*, 113026.
34. Masunga, N., Mmelesi, O. K., Kefeni, K. K., & Mamba, B. B. (2019). Recent advances in copper ferrite nanoparticles and nanocomposite synthesis, magnetic properties, and application in water treatment. *Journal of Environmental Chemical Engineering*, *7*(3), 103179.
35. Rajput, S., Singh, L. P., Pittman Jr, C. U., & Mohan, D. (2017). Lead (Pb^{2+}) and copper (Cu^{2+}) remediation from water using superparamagnetic maghemite (γ-Fe_2O_3) nanoparticles synthesized by Flame Spray Pyrolysis (FSP). *Journal of Colloid and Interface Science*, *492*, 176–190.
36. Khan, K. (2014). Microwave absorption properties of radar absorbing nanosized cobalt ferrites for high-frequency applications. *Journal of Superconductivity and Novel Magnetism*, 27, 453–461.
37. Junaid, M., Khan, M. A., Abubshait, S. A., Akhtar, M. N., Kattan, N. A., Laref, A., & Javed, H. M. A. (2020). Structural, spectral, dielectric, and magnetic properties of indium substituted copper spinel ferrites synthesized via sol-gel technique. *Ceramics International*, *46*(17), 27410–27418.
38. Puli, V. S., Adireddy, S., & Ramana, C. V. (2015). Chemical bonding and magnetic properties of gadolinium (Gd) substituted cobalt ferrite. *Journal of Alloys and Compounds*, *644*, 470–475.
39. Muscas, G., Cobianchi, M., Lascialfari, A., Cannas, C., Musinu, A., Omelyanchik, A., ... Peddis, D. (2019). Magnetic interactions versus magnetic anisotropy in spinel ferrite nanoparticles. *IEEE Magnetics Letters*, *10*, 1–5.
40. Omelyanchik, A., Singh, G., Volochaev, M., Sokolov, A., Rodionova, V., & Peddis, D. (2019). Tunable magnetic properties of Ni-doped $CoFe_2O_4$ nanoparticles prepared by the sol-gel citrate self-combustion method. *Journal of Magnetism and Magnetic Materials*, *476*, 387–391.
41. Anumol, C. N., Chithra, M., Rout, S., & Sahoo, S. C. (2020). Effect of Magnesium substitution on structural and magnetic properties of Nickel ferrite nanoparticles. *Journal of Superconductivity and Novel Magnetism*, *33*, 1611–1617.
42. Manikandan, V., Mirzaei, A., Vigneselvan, S., Kavita, S., Mane, R. S., Kim, S. S., & Chandrasekaran, J. (2019). Role of ruthenium in the dielectric, magnetic properties of nickel ferrite (Ru–$NiFe_2O_4$) nanoparticles and their application in hydrogen sensors. *Acs Omega*, *4*(7), 12919–12926.
43. Raghuvanshi, S., Tiwari, P., Kane, S. N., Avasthi, D. K., Mazaleyrat, F., Tatarchuk, T., & Mironyuk, I. (2019). Dual control on structure and magnetic properties of Mg ferrite: Role of swift heavy ion irradiation. *Journal of Magnetism and Magnetic Materials*, *471*, 521–528.
44. Sagayaraj, R. (2022). A review on structural and magnetic properties of magnesium ferrite nanoparticles. *International Nano Letters*, *12*(4), 345–350.
45. Petrova, E., Kotsikau, D., Pankov, V., & Fahmi, A. (2019). Influence of synthesis methods on structural and magnetic characteristics of Mg–Zn-ferrite nanopowders. *Journal of Magnetism and Magnetic Materials*, *473*, 85–91.

46. Kuru, M., Kuru, T. Ş., Karaca, E., & Bağcı, S. (2020). Dielectric, magnetic, and humidity properties of Mg–Zn–Cr ferrites. *Journal of Alloys and Compounds*, *836*, 155318.
47. Ullah Rather, S., & Lemine, O. M. (2020). Effect of Al doping in zinc ferrite nanoparticles and their structural and magnetic properties. *Journal of Alloys and Compounds*, *812*, 152058.
48. Sathiyamurthy, K., Rajeevgandhi, C., Bharanidharan, S., Sugumar, P., & Subashchandrabose, S. (2020). Electrochemical and magnetic properties of zinc ferrite nanoparticles through chemical co-precipitation method. *Chemical Data Collections*, *28*, 100477.
49. Aisida, S. O., Akpa, P. A., Ahmad, I., Maaza, M., & Ezema, F. I. (2019). Influence of PVA, PVP, and PEG doping on the optical, structural, morphological, and magnetic properties of zinc ferrite nanoparticles produced by thermal method. *Physica B: Condensed Matter*, *571*, 130–136.
50. Ghorbani, H., Eshraghi, M., & Dodaran, A. S. (2022). Structural and magnetic properties of cobalt ferrite nanoparticles doped with cadmium. *Physica B: Condensed Matter*, *634*, 413816.
51. Mahdikhah, V., Ataie, A., Babaei, A., Sheibani, S., Ow-Yang, C. W., & Abkenar, S. K. (2019). Control of structural and magnetic characteristics of cobalt ferrite by post-calcination mechanical milling. *Journal of Physics and Chemistry of Solids*, *134*, 286–294.

5 Spinel and Inverse Ferrites for Drug Delivery

Pulkit Kumar and Harpreet Kaur Channi

5.1 INTRODUCTION

Drug delivery using spinel and inverse spinel ferrites is a groundbreaking field of research that combines materials science and pharmaceuticals to revolutionize drug delivery methods. This innovative approach harnesses the unique properties of spinel and inverse ferrite nanoparticles to enhance drug targeting, release, and efficacy, opening new doors for personalized medicine and improved patient care. Spinel and inverse ferrites are a class of magnetic materials with exceptional characteristics that make them ideal candidates for drug delivery systems. These materials exhibit high surface area, excellent biocompatibility, and tunable magnetic properties, which are pivotal for efficient drug encapsulation and controlled release. They have garnered considerable attention from researchers due to their potential to address critical challenges in conventional drug delivery methods. One of the key advantages of using spinel and inverse ferrite nanoparticles is their ability to carry a wide range of therapeutic agents, including small molecules, proteins, and nucleic acids. Their large surface area-to-volume ratio allows for high drug-loading capacity, ensuring that a significant amount of the therapeutic payload can be transported to the target site. Moreover, these nanoparticles can be functionalized with specific ligands or antibodies, enabling precise targeting of diseased tissues while minimizing collateral damage to healthy cells.

In the ever-evolving landscape of healthcare, the quest for more effective and precise drug delivery methods remains a cornerstone of medical research and innovation. At the intersection of nanotechnology and pharmacology, a revolutionary approach is emerging, poised to redefine how drugs are administered and their impact on patients. The focus here is on a class of materials that have largely flown under the radar of mainstream medicine but holds immense potential for transformative change: spinel and inverse ferrites. This introduction serves as a gateway to explore the world of these remarkable materials and their applications in drug delivery, a journey that unfolds at the nexus of science, medicine, and patient care.

 DOI: 10.1201/9781032665658-5

5.2 FERRITES: THE UNASSUMING MARVELS

Ferrites, with their crystalline structures, possess a unique combination of electrical, magnetic, and optical properties that have been harnessed in numerous technological applications, from electronic components to magnetic storage devices. However, it is their potential in the realm of drug delivery that is capturing the imagination of scientists and healthcare professionals alike. To understand this potential, it is first essential to delve into the world of ferrites, the key players in this evolving narrative.

Ferrites are compounds consisting of iron (Fe^{3+}) and other metal ions, such as manganese, zinc, or nickel, arranged in a specific crystal lattice structure. Spinel and inverse ferrites, in particular, have garnered attention for their magnetic properties, which can be tailored to various biomedical applications. The term "spinel" refers to a specific crystal structure, while "inverse ferrites" encompass a range of ferrite materials with unique magnetic behaviors.

5.3 SPINEL FERRITES: THE MAGNETIC PIONEERS

Spinel ferrites are renowned for their magnetic properties. These materials, which include magnetite (Fe_3O_4) and maghemite (γ-Fe_2O_3), are ferrimagnetic, meaning they possess both ferromagnetic and anti-ferromagnetic properties. This duality makes them highly attractive for drug delivery applications. The magic lies in their capacity to respond to external magnetic fields.

When engineered into nanoparticles, spinel ferrites can act as carriers for therapeutic agents, such as drugs or genes. These nanoparticles, when administered into the body, can be precisely guided to specific target sites using external magnetic fields. This magnetic targeting enables a level of precision previously unattainable in drug delivery, as it reduces off-target effects and enhances drug concentrations at the desired location. This holds tremendous promise in the treatment of various diseases, especially in the realm of cancer therapy.

5.4 INVERSE FERRITES: OVERCOMING BIOLOGICAL BARRIER

While spinel ferrites are magnetic pioneers, inverse ferrites offer a unique set of characteristics that allow them to overcome significant barriers within the human body, particularly the blood–brain barrier (BBB). The blood–brain barrier, a protective shield for the central nervous system, restricts the passage of most substances from the bloodstream into the brain and spinal cord. This presents a substantial challenge in delivering drugs to the central nervous system for the treatment of neurological disorders.

Inverse ferrites, however, have demonstrated the remarkable ability to bypass the BBB. These nanoparticles can be designed to be stealthy, evading the vigilant defense mechanisms of the barrier. This attribute is of paramount significance for treating conditions such as Alzheimer's disease, Parkinson's disease, brain tumors, and neurodegenerative disorders, where delivering therapeutic agents to the central nervous system is paramount. The magnetic properties of spinel and inverse ferrites are the linchpin of their application in drug delivery. These properties can be tailored and optimized for specific medical purposes, rendering them versatile tools in the healthcare arsenal. First and foremost, the responsiveness of these nanoparticles to external magnetic fields allows for unparalleled precision in drug targeting. Magnetic fields can be applied externally to guide drug-loaded nanoparticles directly to the site of action, reducing collateral damage to healthy tissues and increasing the therapeutic impact on the intended target. This magnetic guidance ensures that drugs reach their destinations with remarkable accuracy.

Second, the magnetic properties of ferrites facilitate controlled drug release. By applying magnetic fields or altering the conditions within the body, it is possible to trigger the release of therapeutic agents from the nanoparticles at precisely the right moment. This controlled release not only enhances treatment efficacy but also minimizes the risk of overexposure to drugs, which can lead to side effects.

5.5 LITERATURE REVIEW

Chen, Liu et al. (2018) have presented biomass-composited inverse opal particles with a series of advanced features for drug delivery and wound healing. X-ray diffraction (XRD) and vibrational spectroscopy techniques were employed to study the structure, cation distribution, and order in the Co^{2+}-substituted nickel ferrite (NCFO) samples [1]. Park et al. (2019) presented a short review of the state-of-the-art progress of developments of graphene-based nano spinel ferrites (GNSFs), as well as their potential application for removing several recalcitrant contaminants including organic dyes, antibiotics, and heavy metal ions [2]. Salavati-Niasari et al. (2019) have summarized the theory of magnetic nanoparticles and magnetic spinel ferrite properties, remarkable approaches in magnetic liposomes, magnetic polymeric nanoparticles, magnetic resonance imaging (MRI), hyperthermia, and especially magnetic drug delivery systems, which have recently developed in the field of magnetic nanoparticles and their medicinal applications [3]. Somvanshi et al. (2020) analyzed the room temperature X-ray diffraction (XRD) patterns that showed the formation of cubic spinel structure with a single phase in the Zn–Mg ferrites [4]. Sharifianjazi et al. (2020) reviewed the $CoFe_2O_4$ nanoparticles doped with various elements and their applications in various fields [5]. Peng et al. (2021) proposed future recommendations and challenges related to spinel ferrites (SFs) and spinel ferrite-based composites (SFCs) catalysts, heterogeneous SFs/SFCs-PMS systems, and their possible applications in wastewater purification. These composites often exhibit efficiency in fulfilling coveted magnetic and dielectric characteristics [6]. Houbi et al. (2021) provided a brief presentation of ferrites and among them are spinel ferrites and hexagonal ferrites [7].

Cruickshank et al. (2020) have revealed an invention on modified Ni-Zn ferrites for radiofrequency applications and also described spinel-based oxides containing magnesium, aluminum, and titanium and methods of forming articles having the same [8]. Chen et al. (2023) described magneto-dielectric materials, methods of making, and uses thereof [9]. Zhang and Chen (2021) have described nanocrystalline cobalt-doped nickel ferrite particles, method of manufacture, and uses [10]. Hessian and Hanafy described nickel-zinc ferrites and methods for preparing the same using fine iron oxide and bag house dust [11].

5.5.1 Ferrite Nanoparticles: Structure and Properties

"Ferrite Nanoparticles: Structure and Properties" is a fascinating topic that delves into the intricate world of nanomaterials with remarkable magnetic properties and diverse applications. Ferrite nanoparticles, a subset of these materials, have garnered considerable attention in recent years due to their unique structure and a wide range of properties that make them highly valuable in various fields, including electronics, medicine, and environmental remediation. At their core, ferrite nanoparticles are composed of iron oxide (Fe_3O_4 or Fe_2O_3) and exhibit a crystal structure known as spinel or inverse spinel. This structure consists of oxygen ions forming a face-centered cubic (FCC) lattice, with iron ions occupying either octahedral or tetrahedral sites within the lattice. This arrangement results in a magnetic moment, making ferrite nanoparticles ferromagnetic or ferrimagnetic, depending on their composition and size [12].

One of the defining features of ferrite nanoparticles is their size, typically ranging from 1 to 100 nm. This nanoscale dimension gives rise to several unique properties. For instance, as nanoparticles become smaller, their surface area relative to their volume increases significantly. This high surface area is a double-edged sword – it allows for enhanced reactivity and unique surface-dependent properties but also raises challenges related to stability and control during synthesis. The magnetic properties of ferrite nanoparticles are a result of the alignment of magnetic moments within the crystal lattice. In ferromagnetic ferrites, like magnetite (Fe_3O_4), the magnetic moments of iron ions align in the same direction, resulting in a net magnetic moment. In ferrimagnetic ferrites, like maghemite (γ-Fe_2O_3), there is partial antiparallel alignment, leading to a lower net magnetic moment. These magnetic properties are highly desirable and underpin numerous technological applications [13].

One of the most prominent applications of ferrite nanoparticles is in magnetic data storage. Their ability to exhibit strong magnetization while maintaining a small size makes them ideal candidates for high-density storage media, such as hard drives. The data bits are stored as magnetic domains on the surface of nanoparticles, with their orientation representing binary information. This technology has revolutionized data storage, enabling the capacity to store vast amounts of information in a compact form. In addition to data storage, ferrite nanoparticles find extensive use in the field of biomedical applications. Their biocompatibility, along with their magnetic properties, allows for targeted drug delivery and magnetic resonance imaging (MRI). Functionalized ferrite nanoparticles can be coated with biocompatible materials and specific ligands to enable site-specific drug delivery. When an external magnetic field is applied, these nanoparticles can be guided to the target site, improving drug concentration at the desired location and reducing side effects. Similarly, in MRI, ferrite nanoparticles can serve as contrast agents, enhancing the visibility of tissues and providing valuable diagnostic information [14].

Furthermore, the magnetic properties of ferrite nanoparticles are exploited in hyperthermia therapy. When exposed to an alternating magnetic field, these nanoparticles generate heat through hysteresis losses. This heat can be precisely controlled and utilized to selectively destroy cancer cells or target diseased tissues, offering a promising avenue for cancer treatment. Beyond biomedical applications, ferrite nanoparticles play a crucial role in environmental remediation. They can be used to remove heavy metals and other contaminants from polluted water sources. The nanoparticles adsorb these contaminants onto their surfaces, effectively purifying the water. This application is vital for addressing environmental challenges and ensuring access to clean water resources [15].

The electronic industry also benefits from ferrite nanoparticles. They are used in the manufacturing of inductors and transformers due to their high magnetic permeability and low eddy current losses. These properties are essential for the efficient operation of electronic circuits and power supply systems. Despite their many advantages, the synthesis and stabilization of ferrite nanoparticles can be challenging. Controlling the size, shape, and surface chemistry of these nanoparticles is crucial in tailoring their properties for specific applications. Various synthesis methods, including sol–gel, co-precipitation, and hydrothermal processes, have been developed to achieve this control. Moreover, surface modification with organic molecules or polymers is often necessary to enhance stability and prevent agglomeration [16].

5.5.1.1 Crystal Structure of Spinel Ferrites

The crystal structure of spinel ferrites is a fascinating and well-studied topic in materials science. Spinel ferrites belong to a class of magnetic materials with a distinct crystal structure known as the spinel structure. The general formula for spinel ferrites is AB_2O_4, where A and B represent different metal ions, typically with A as a divalent metal ion (e.g., Fe^{2+}, Zn^{2+}, Ni^{2+}) and B as a trivalent metal ion (e.g., Fe^{3+}, Al^{3+}, Cr^{3+}). The oxygen ions (O_2-) complete the crystal structure, forming an intricate lattice arrangement as shown in Figure 5.1 [17].

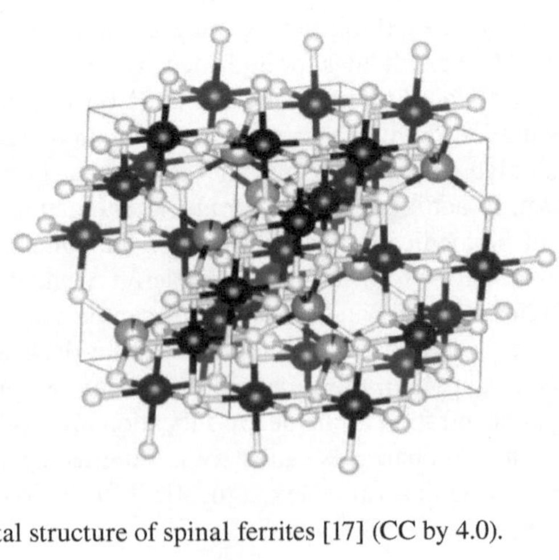

FIGURE 5.1 Crystal structure of spinal ferrites [17] (CC by 4.0).

5.5.1.2 The Unique Inverse Spinel Structure

The inverse spinel structure is a distinct crystal arrangement found in specific materials, setting them apart from those with a regular spinel structure. It is characterized by a unique distribution of metal cations within the crystal lattice. In the inverse spinel, two-thirds of the tetrahedral sites are occupied by one type of cation (usually a divalent metal ion), while one-third of the octahedral sites are filled by a mixture of two different cations (typically a trivalent and a divalent metal ion) (Figure 5.2). This unconventional cation distribution leads to exceptional properties. Magnetite (Fe_3O_4) is a well-known example of a material with an inverse spinel structure. Its distinctive arrangement of iron cations results in robust magnetic properties, making it valuable in applications such as magnetic storage media and magnetic resonance imaging (MRI). The inverse spinel structure exemplifies how variations in crystal arrangement can give rise to materials with unique and valuable characteristics [18].

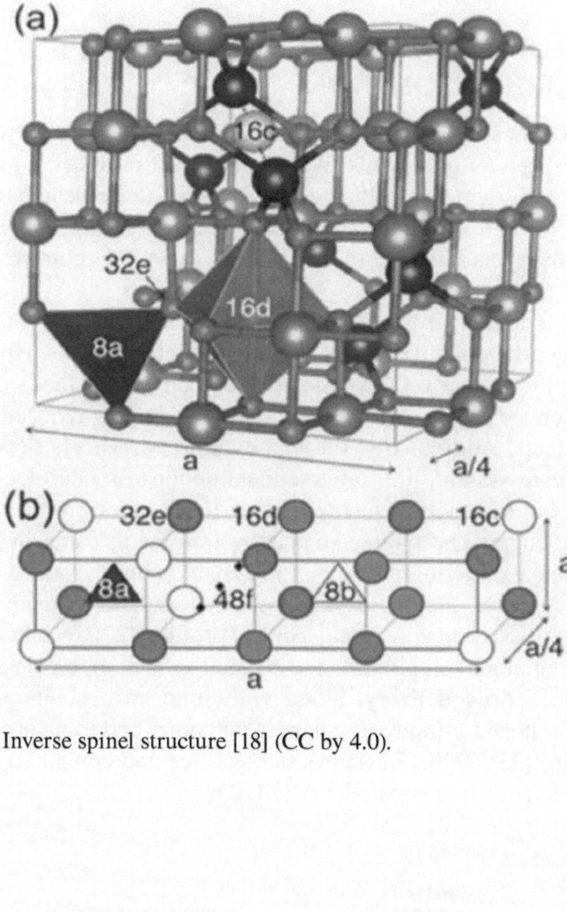

FIGURE 5.2 Inverse spinel structure [18] (CC by 4.0).

5.6 MAGNETIC PROPERTIES OF SPINEL AND INVERSE SPINEL FERRITES

Magnetic properties are fundamental characteristics of materials that arise from the behavior of their atomic or molecular magnetic moments. These properties play a significant role in a wide range of scientific, technological, and industrial applications. Ferromagnetism, paramagnetism, and diamagnetism are the primary magnetic properties of materials. Ferromagnetic materials, like iron and nickel, exhibit strong, permanent magnetization due to the alignment of atomic magnetic moments. This property is pivotal in applications such as data storage, electromagnets, and electric motors. Paramagnetic materials possess weak, temporary magnetization when exposed to an external magnetic field, making them valuable in magnetic resonance imaging (MRI) and materials characterization techniques. Diamagnetic materials, on the other hand, are weakly repelled by magnetic fields and include most non-magnetic substances. These magnetic properties have significant practical implications. They enable the development of technologies ranging from electrical generators and transformers to medical diagnostics and magnetic levitation trains. Understanding and harnessing these properties are essential for innovations in science and engineering [19].

5.7 DRUG DELIVERY: CURRENT CHALLENGES AND EMERGING SOLUTIONS

Drug delivery is a critical component of modern healthcare, as it directly impacts the effectiveness and safety of treatments. While significant progress has been made in drug delivery systems, there are still several challenges that researchers and healthcare professionals are actively addressing. Here, we explore some of the current challenges in drug delivery and the emerging solutions that offer promise for the future of medicine [20].

Drug delivery is a critical aspect of modern healthcare, with the potential to significantly impact the effectiveness of therapeutic interventions. However, it also presents a range of challenges. Conventional drug delivery methods often result in suboptimal drug concentrations at the target site, leading to reduced efficacy and increased side effects. Additionally, the blood–brain barrier restricts drug access to the central nervous system, limiting treatment options for neurological disorders [21]. Emerging solutions, such as ferrite nanoparticles, offer promising avenues to address these challenges. These innovative delivery systems can provide controlled drug release, enhance targeting precision, and enable non-invasive routes for drug administration. Moreover, they hold potential for personalized medicine, offering tailored treatments based on individual patient profiles [22]. While challenges in terms of safety, regulatory compliance, and scalability remain, the ongoing research and development in drug delivery, along with innovative solutions like ferrite nanoparticles, offer hope for improving patient outcomes and transforming the landscape of healthcare [23]. Table 5.1 shows the features and comparative analysis of spinal vs. inverse ferrites for drug delivery [24, 25].

TABLE. 5.1
Spinel vs. Inverse Spinel Ferrites for Drug Delivery

Feature	Spinel Ferrite	Inverse Spinel Ferrite
Cationdistribution	A^{2+} $[B^{2+}(1-\delta)Fe^{3}+\delta]O_4$	A^{2+} $[B^{2+}\delta\ Fe^{3+}(1-\delta)]O_4$
Magnetic properties	Vary depending on composition, often superparamagnetic or ferrimagnetic	Generally ferrimagnetic due to larger Fe^{3+} content in octahedral sites
Drug-loading capacity	Generally lower due to fewer accessible sites for drug attachment	Higher due to more available octahedral sites
Drug release mechanisms	Similar to inverse spinel ferrites, including pH, magnetic field, and enzymatic triggers	May offer additional release mechanisms due to cation exchange in octahedral sites
Biocompatibility	Generally good, but may depend on specific composition	May vary depending on the A^+ cation, and some require additional coatings for biocompatibility
Imaging capabilities	Some compositions exhibit superparamagnetism, allowing for MRI tracking	Limited imaging capabilities due to lower Fe^{3+} content
Advantages	Often simpler and cheaper to synthesize, broader range of magnetic properties	Higher drug-loading capacity, potential for additional release mechanisms
Disadvantages	Lower drug-loading capacity, limited imaging capabilities	May require additional biocompatibility coatings, potential for cation exchange affecting stability
Examples	$MnFe_2O_4$ (manganese ferrite), $CoFe_2O_4$ (cobalt ferrite)	$NiFe_2O_4$ (nickel ferrite), $MgFe_2O_4$ (magnesium ferrite)

Traditional drug delivery methods often fall short of providing the desired therapeutic outcomes due to issues like inefficient drug targeting, systemic exposure leading to adverse side effects, and difficulties in crossing biological barriers, such as the blood–brain barrier. This presents a pressing need for innovative solutions to enhance drug delivery precision and efficiency [26]. Emerging solutions, including the use of ferrite nanoparticles, show significant promise in overcoming these challenges. Ferrite nanoparticles can be engineered to offer controlled and localized drug release, allowing for precise dosing and minimizing off-target effects. Their magnetic properties enable magnetic targeting, guiding drug-loaded nanoparticles to specific sites within the body, potentially revolutionizing the treatment of diseases like cancer, where targeted drug delivery is crucial. Furthermore, ferrite nanoparticles have the potential to bypass biological barriers, such as the blood–brain barrier, thus expanding treatment options for conditions affecting the central nervous system [27].

One of the most exciting aspects of these emerging drug delivery technologies is their potential for personalized medicine. By tailoring drug delivery to individual patient profiles, it becomes possible to optimize treatment efficacy while minimizing side effects, making healthcare more patient-centered [28]. However, as with any innovative medical technology, challenges persist. Safety and biocompatibility assessments must be thorough to ensure that ferrite nanoparticles are free from harm and interact safely with biological systems. Regulatory compliance and scalability are essential considerations for the widespread adoption of these technologies, and long-term efficacy and ethical implications are areas that warrant ongoing exploration. In conclusion, the field of drug delivery is at an exciting crossroads, with emerging solutions like ferrite nanoparticles offering the potential to address the limitations of conventional methods and revolutionize healthcare. Ongoing research, development, and regulatory collaboration will play pivotal roles in bringing these innovative drug delivery technologies to the forefront of medical practice, offering the promise of more effective and personalized treatments for patients worldwide [29].

5.8 APPLICATIONS OF FERRITE-BASED DRUG DELIVERY SYSTEMS

Ferrite-based drug delivery systems, harnessing the exceptional properties of ferrite nanoparticles, have found diverse applications in the field of medicine. These systems enable precise and controlled drug delivery, significantly improving therapeutic outcomes while minimizing the unwanted side effects typically associated with conventional treatments. Notably, they have made significant strides in various medical areas. In cancer therapy, for instance, these systems have redefined the landscape by loading magnetic nanoparticles with chemotherapy drugs and precisely guiding them to tumor sites using external magnetic fields, thus reducing collateral damage to healthy tissues [30]. Additionally, in central nervous system (CNS) drug delivery, they have addressed the challenge of the blood–brain barrier (BBB), allowing for the targeted treatment of neurological disorders. Furthermore, in the management of inflammatory and autoimmune diseases, ferrite-based drug delivery systems provide the ability to target and deliver anti-inflammatory or immunosuppressive drugs precisely to sites of inflammation or autoimmune reactions.

This approach minimizes systemic side effects and ensures controlled drug release, thereby enhancing treatment efficacy. These versatile systems are continually advancing, offering the promise of improving patient outcomes across a wide spectrum of medical fields. Ferrite-based drug delivery systems offer a diverse range of applications in the medical and pharmaceutical fields. These applications are characterized by their potential to significantly enhance the precision, efficiency, and safety of drug delivery. In the realm of oncology, these systems have proven invaluable [31]. By loading magnetic nanoparticles with chemotherapy drugs and utilizing external magnetic fields, they enable targeted drug delivery to cancerous cells while sparing healthy tissues. This targeted approach reduces the debilitating side effects often associated with traditional chemotherapy. In neurology, the challenges posed by the blood-brain barrier have limited the treatment options for neurological disorders. Ferrite nanoparticles can be designed to bypass this barrier, allowing for the delivery of therapeutic agents to the central nervous system. This opens up new avenues for treating conditions like Alzheimer's disease, Parkinson's disease, and brain tumors, which were previously difficult to access [32].

Ferrite-based drug delivery systems have shown tremendous promise in various medical and pharmaceutical applications. These applications are characterized by their ability to provide highly targeted and controlled drug delivery, thereby revolutionizing conventional treatment approaches. In the realm of oncology, these systems are a beacon of hope for cancer patients. By encapsulating chemotherapy drugs within magnetic nanoparticles and using external magnetic fields, ferrite-based systems ensure that the drugs are precisely delivered to the tumor site [33]. This targeted therapy significantly reduces the damage to healthy tissues, alleviating the often debilitating side effects associated with chemotherapy. For central nervous system (CNS) disorders, such as brain tumors or neurodegenerative diseases, the blood–brain barrier (BBB) has long posed a formidable challenge for drug delivery. Ferrite nanoparticles can be engineered to cross the BBB, allowing for the delivery of therapeutic agents directly to the CNS. This approach opens up new possibilities for treatments that were previously challenging to administer effectively. Furthermore, in the management of inflammatory and autoimmune diseases, ferrite-based drug delivery systems hold great potential. They enable the precise delivery of anti-inflammatory or immunosuppressive drugs directly to the sites of inflammation or autoimmune reactions. This targeted delivery minimizes systemic side effects, offering patients a more tolerable and effective treatment option [34].

Additionally, the versatility of ferrite nanoparticles extends beyond therapeutics. They are utilized as contrast agents in medical imaging, particularly in magnetic

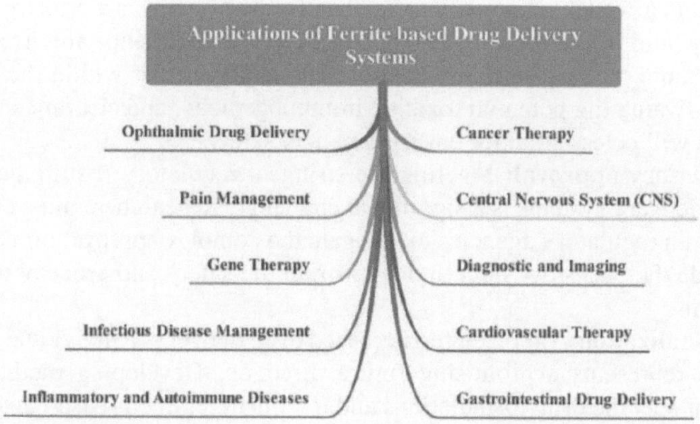

FIGURE 5.3 Applications of ferrite-based drug delivery systems.

resonance imaging (MRI). This enables non-invasive and real-time monitoring of disease progression and the distribution of nanoparticles within the body during therapy. In summary, the applications of ferrite-based drug delivery systems are wide-ranging and transformative, significantly improving treatment outcomes, minimizing side effects, and advancing the field of personalized medicine. The ongoing research and development in this area offer promising prospects for enhancing patient care and quality of life across various medical disciplines. Ferrite-based drug delivery systems have gained recognition for their unique properties and versatility in various pharmaceutical applications. These systems, often utilizing ferrite nanoparticles, offer several promising applications in the field of drug delivery. Different applications of drug delivery systems are shown in Figure 5.3.

The field of spinel and inverse ferrites for drug delivery holds immense promise, but it also faces several future directions and challenges that researchers and scientists need to address

- **Clinical translation:** One of the primary future directions is the clinical translation of these innovative drug delivery systems. While significant progress has been made in preclinical studies, transitioning from lab-based research to human trials and regulatory approval remains a formidable challenge. Rigorous safety and efficacy evaluations will be necessary to demonstrate the clinical viability of these systems.
- **Targeting precision:** Enhancing the precision of drug targeting remains an ongoing challenge. Fine-tuning the surface functionalization of nanoparticles to ensure specific binding to target cells while minimizing off-target effects is crucial. Achieving consistent and reliable targeting across diverse patient populations is a goal that requires further research and optimization.
- **Scalability and cost-efficiency:** Developing scalable and cost-effective manufacturing processes for ferrite-based drug delivery systems is essential for their widespread adoption. Ensuring that these technologies are accessible and affordable for patients and healthcare systems is a priority [35].
- **Safety and toxicity:** Comprehensive safety assessments are needed to understand the long-term effects of ferrite nanoparticles within the human body. Evaluating potential toxicity, immunogenicity, and clearance mechanisms will be essential for ensuring patient safety.
- **Regulatory approval:** Meeting the stringent regulatory requirements for drug delivery systems is a significant challenge. Researchers must collaborate with regulatory agencies to navigate the complex approval process and provide the necessary data to demonstrate the safety and efficacy of these systems.
- **Personalization:** Tailoring ferrite-based drug delivery to individual patient needs represents a promising future direction. Developing methods for patient-specific drug formulations and treatment regimens based on genetic and physiological factors can optimize therapeutic outcomes.
- **Multimodal approaches:** Integrating ferrite-based drug delivery with other therapeutic modalities, such as immunotherapy or gene therapy, holds great

potential. However, combining multiple treatments introduces complexities in dosing, timing, and synergistic effects that require careful investigation.

- **Ethical and regulatory considerations:** As with any emerging technology, ethical considerations surrounding the use of ferrite-based drug delivery must be addressed. Researchers and policymakers should engage in discussions regarding patient consent, privacy, and equitable access to these therapies.
- **Long-term efficacy:** Assessing the long-term efficacy and potential development of resistance in response to ferrite-based drug delivery systems is a critical challenge. Monitoring and adapting treatments over extended periods will be necessary [36].

5.9 PRACTICAL APPLICATIONS OF FERRITES

5.9.1 Ferrite Nanoparticles in Targeted Chemotherapy

In this case study, ferrite nanoparticles were employed in targeted chemotherapy to enhance the precision and efficacy of cancer treatment. Researchers developed ferrite-based drug delivery systems functionalized with tumor-specific ligands. These nanoparticles were loaded with a potent chemotherapeutic drug and administered to patients with aggressive brain tumors. The magnetic properties of ferrite nanoparticles allowed for magnetic targeting, guiding the drug-loaded nanoparticles specifically to the tumor site. This targeted approach minimized drug exposure to healthy brain tissue, reducing side effects. The results of this study demonstrated improved therapeutic outcomes, with enhanced tumor regression and prolonged patient survival compared to traditional systemic chemotherapy. This case study highlights the potential of ferrite nanoparticles to revolutionize cancer therapy by delivering drugs precisely to tumor sites while sparing healthy tissue [37, 38].

5.9.2 Magnetic Hyperthermia for Cancer Treatment

In this case study, magnetic hyperthermia was explored as a novel approach for cancer treatment using ferrite nanoparticles. Researchers designed magnetic nanoparticles with specific heating properties when exposed to an alternating magnetic field. These nanoparticles were loaded with heat-sensitive drugs and injected into patients with inoperable liver tumors. When subjected to an external alternating magnetic field, the nanoparticles generated localized heat within the tumor, triggering the release of the enclosed drugs. This combination of hyperthermia and drug release led to effective tumor ablation while minimizing damage to surrounding healthy tissues. The case study demonstrated the potential of magnetic hyperthermia using ferrite nanoparticles as a minimally invasive and highly precise cancer treatment modality [39, 40].

5.9.3 RNA Delivery Using Ferrite Nanoparticles

In this case study, researchers explored the use of ferrite nanoparticles for the targeted delivery of RNA-based therapies. The goal was to develop a platform for delivering small interfering RNA (siRNA) to inhibit the expression of specific genes implicated in cancer. Ferrite nanoparticles were functionalized with siRNA molecules and modified with ligands targeting cancer cells. When administered to patients with aggressive breast cancer, these nanoparticles effectively delivered siRNA to the tumor cells, leading to the selective suppression of oncogenes responsible for tumor growth. This innovative approach demonstrated the potential of ferrite nanoparticles in RNA-based cancer therapy, offering a precise and customizable way to modulate gene expression in cancer cells and halt tumor progression [41].

These case studies illustrate the versatility and promise of ferrite nanoparticles in advancing targeted and personalized cancer treatments. Whether through enhanced drug delivery, magnetic hyperthermia, or RNA-based therapies, ferrite nanoparticles have the potential to significantly improve therapeutic outcomes while minimizing the side effects often associated with conventional cancer treatments [32].

5.10 SAFETY AND BIOCOMPATIBILITY CONSIDERATIONS

Safety and biocompatibility considerations are paramount when developing and utilizing ferrite nanoparticles for drug delivery and other medical applications. These considerations are essential to ensure that ferrite nanoparticles are not only effective in their intended therapeutic role but also do not harm the human body or negatively impact biological systems. Firstly, comprehensive toxicity assessments are fundamental. Researchers must rigorously evaluate the potential adverse effects of ferrite nanoparticles on cellular and tissue levels. This includes examining acute and chronic toxicity, investigating the potential for cellular damage, and assessing their impact on different organ systems. These studies are typically conducted in vitro, using cell cultures, and in vivo, through animal studies [32].

Understanding the long-term effects of ferrite nanoparticles is crucial. Researchers need to investigate whether these nanoparticles accumulate in specific organs or tissues over time and whether such accumulation has any adverse consequences. Long-term safety data are vital for ensuring the continued well-being of patients who may receive treatments involving ferrite nanoparticles. Immunogenicity is another critical aspect to consider. Researchers need to assess whether ferrite nanoparticles trigger immune responses in the body. This involves evaluating whether the nanoparticles induce inflammation, activate immune cells, or provoke allergic reactions. Such immune responses can have significant implications for patient safety and overall treatment effectiveness [42].

Additionally, researchers must investigate how the body clears ferrite nanoparticles. Understanding the clearance mechanisms helps determine whether the nanoparticles are metabolized and excreted from the body or whether they persist in the tissues. Biodistribution studies are essential for identifying where these materials accumulate within the body, shedding light on potential off-target effects. Surface modifications play a vital role in enhancing biocompatibility. By functionalizing the nanoparticles with biocompatible coatings or polymers, researchers can reduce their interaction with biological molecules and improve their safety profile. This can minimize the risk of adverse reactions when ferrite nanoparticles come into contact with biological fluids or tissues [43].

Hemocompatibility is a crucial consideration if ferrite nanoparticles are to interact with blood. Assessing their impact on clotting, platelet activation, and red blood cell integrity is essential to avoid unintended consequences when using these materials in applications that involve blood contact. Furthermore, compliance with regulatory guidelines and safety standards is paramount. Researchers and developers must collaborate closely with regulatory authorities to ensure that their ferrite-based drug delivery systems meet the required safety and efficacy standards before advancing to clinical trials and commercialization. Finally, ethical considerations, such as patient consent, privacy protection, and equitable access to innovative therapies, should be an integral part of the safety and biocompatibility assessment process, ensuring that the rights of patients and well-being are always upheld [44].

5.10.1 Biocompatibility of Ferrite Nanoparticles

The biocompatibility of ferrite nanoparticles refers to their ability to interact with biological systems without causing harm or adverse effects. It involves assessing whether these nanoparticles are safe for use in medical applications. Evaluation includes cell viability tests to ensure they do not harm cells, examinations for inflammatory responses, checks for compatibility with blood, and studies on tissue and organ compatibility. Understanding how these nanoparticles are distributed and cleared from the body and assessing their long-term effects are also essential. Achieving biocompatibility is crucial for the safe and effective use of ferrite nanoparticles in areas like drug delivery, diagnostics, and treatments [45].

5.10.2 Toxicological Assessment

Toxicological assessment is a systematic evaluation process used to determine the potential toxicity or harmful effects of substances, such as chemicals, drugs, or nanoparticles, on biological systems. It involves conducting experiments and studies to assess the adverse impacts of these substances on living organisms, including humans. Toxicological assessments examine factors like acute and chronic toxicity, carcinogenicity, genotoxicity, and teratogenicity. The goal is to identify safe exposure levels, establish regulatory guidelines, and ensure the safety of products and materials, especially those used in pharmaceuticals, chemicals, and environmental contexts. These assessments inform risk management decisions and contribute to safeguarding human and environmental health [46].

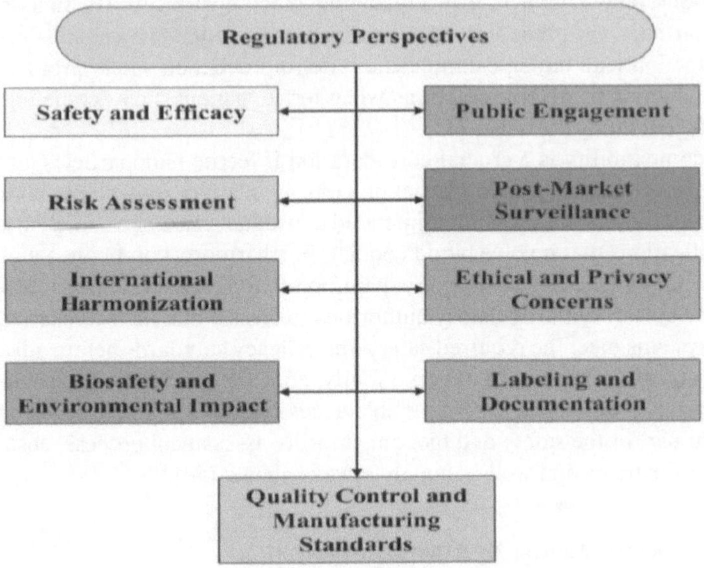

FIGURE 5.4 Regulatory perspectives on the use of ferrite nanoparticles.

5.10.3 Regulatory Perspectives

Regulatory perspectives on the use of ferrite nanoparticles, particularly in drug delivery and medical applications, are essential to ensure the safety, efficacy, and ethical use of these innovative materials (Figure 5.4). Regulatory agencies around the world play a crucial role in evaluating and overseeing the development, testing, and commercialization of products that incorporate ferrite nanoparticles. Here are some key regulatory perspectives and considerations:

- **Safety and efficacy:** Regulatory agencies, such as the US Food and Drug Administration (FDA), the European Medicines Agency (EMA), and others, require rigorous safety and efficacy assessments for products containing ferrite nanoparticles. These assessments involve evaluating the potential risks and benefits of the technology, ensuring that it meets established safety standards, and demonstrating its effectiveness in clinical settings [32].
- **Risk assessment:** Comprehensive risk assessments are conducted to identify and mitigate potential risks associated with ferrite nanoparticles. This includes evaluating toxicity, immunogenicity, long-term effects, and other safety concerns. Data from preclinical and clinical studies are analyzed to determine the level of risk and appropriate risk mitigation strategies.
- **Quality control and manufacturing standards:** Regulatory agencies require adherence to strict quality control and manufacturing standards for products containing ferrite nanoparticles. Manufacturers must demonstrate

consistency in production processes, product quality, and safety. Regulatory bodies may inspect manufacturing facilities to ensure compliance with Good Manufacturing Practices (GMP) [47].

- **Biosafety and environmental impact:** Beyond human health considerations, regulatory perspectives also encompass the potential environmental impact of ferrite nanoparticles. Evaluations may include the assessment of nanoparticle release into ecosystems and their impact on aquatic life, soil, and air quality.
- **Ethical and privacy concerns:** Regulatory bodies address ethical and privacy concerns related to patient consent, data protection, and equitable access to innovative therapies. Ensuring that patients are well-informed and protected throughout the development and use of ferrite nanoparticles is a key consideration [48].
- **Labeling and documentation:** Clear labeling and documentation of products containing ferrite nanoparticles are essential for regulatory compliance. This includes providing information on nanoparticle content, usage instructions, potential risks, and safety precautions.
- **International harmonization:** Regulatory agencies often collaborate internationally to harmonize standards and guidelines for the use of nanomaterials like ferrite nanoparticles. This helps streamline the regulatory process, facilitate global trade, and ensure consistent safety evaluations.
- **Post-market surveillance:** After approval, regulatory agencies may require post-market surveillance to monitor the long-term safety and performance of products incorporating ferrite nanoparticles. Adverse events and patient outcomes are closely monitored to identify any emerging safety concerns.
- **Public engagement:** Regulatory perspectives also involve engaging with the public, healthcare professionals, and stakeholders to gather input, address concerns, and foster transparency in decision-making processes [49, 50].

5.11 CONCLUSION

In conclusion, ferrite nanoparticles represent a promising frontier in the realm of drug delivery and medical applications. Their unique properties, such as magnetic targeting, controlled drug release, and biocompatibility, offer innovative solutions to address the limitations of conventional therapies. From targeted cancer treatment to central nervous system drug delivery and RNA-based therapies, ferrite nanoparticles have demonstrated their potential to revolutionize healthcare.

However, realizing this potential requires careful consideration of various factors. Safety and biocompatibility assessments are paramount to ensure that these nanoparticles do not pose harm to patients. Regulatory perspectives play a crucial role in setting standards, evaluating risks, and ensuring ethical use. Furthermore, addressing challenges like scalability, long-term efficacy, and personalized medicine will be pivotal in advancing ferrite nanoparticle-based therapies.

As research continues to evolve, the future holds great promise for ferrite nanoparticles, with the potential to enhance patient outcomes, minimize side effects, and usher in a new era of precision medicine. Collaborative efforts among researchers, regulatory agencies, and healthcare professionals will be instrumental in translating these innovations into tangible benefits for patients worldwide. The journey toward harnessing the full potential of ferrite nanoparticles in healthcare is ongoing, offering hope for improved treatments and a better quality of life for those facing complex medical challenges.

REFERENCES

1. Chen, C., et al., Multifunctional chitosan inverse opal particles for wound healing. *ACS Nano*, 2018. **12**(10): p. 10493–10500.
2. Park, C.M., et al., Potential utility of graphene-based nano spinel ferrites as adsorbent and photocatalyst for removing organic/inorganic contaminants from aqueous solutions: A mini review. *Chemosphere*, 2019. **221**: p. 392–402.
3. Amiri, M., M. Salavati-Niasari, and A. Akbari, Magnetic nanocarriers: Evolution of spinel ferrites for medical applications. *Advances in Colloid and Interface Science*, 2019. **265**: p. 29–44.
4. Somvanshi, S.B., et al., Influential diamagnetic magnesium (Mg2+) ion substitution in nano-spinel zinc ferrite (ZnFe2O4): Thermal, structural, spectral, optical and physisorption analysis. *Ceramics International*, 2020. **46**(7): p. 8640–8650.
5. Sharifianjazi, F., et al., Magnetic $CoFe_2O_4$ nanoparticles doped with metal ions: A review. *Ceramics International*, 2020. **46**(11): p. 18391–18412.
6. Peng, Y., et al., Activation of peroxymonosulfate (PMS) by spinel ferrite and their composites in degradation of organic pollutants: A review. *Chemical Engineering Journal*, 2021. **414**: p. 128800.
7. Houbi, A., et al., Microwave absorbing properties of ferrites and their composites: A review. *Journal of Magnetism and Magnetic Materials*, 2021. **529**: p. 167839.
8. Hill, M.D., D.B. Cruickshank, and K.M. Anderson, *Modified Ni—Zn ferrites for radiofrequency applications*. 2019, Google Patents.
9. Chen, Y., R.C. Daigle, and L. Zhang, *Magneto-dielectric materials, methods of making, and uses thereof*. 2023, Google Patents.
10. Zhang, L. and Y. Chen, *Nanocrystalline cobalt doped nickel ferrite particles, method of manufacture, and uses thereof*. 2021, Google Patents.
11. Hessien, M.M. and H. Hanafy, *Nickel-zinc ferrites and methods for preparing same using fine iron oxide and bag house dust*. 2016, Google Patents.
12. Powar, R.R., et al., Effect of zinc substitution on magnesium ferrite nanoparticles: Structural, electrical, magnetic, and gas-sensing properties. *Materials Science and Engineering: B*, 2020. **262**: p. 114776.
13. Jabbar, R., S.H. Sabeeh, and A.M. Hameed, Structural, dielectric and magnetic properties of Mn^{+2} doped cobalt ferrite nanoparticles. *Journal of Magnetism and Magnetic Materials*, 2020. **494**: p. 165726.
14. Swatsitang, E., et al., Characterization and magnetic properties of cobalt ferrite nanoparticles. *Journal of Alloys and Compounds*, 2016. **664**: p. 792–797.
15. Andhare, D.D., et al., Effect of Zn doping on structural, magnetic and optical properties of cobalt ferrite nanoparticles synthesized via. *Co-precipitation Method. Physica B: Condensed Matter*, 2020. **583**: p. 412051.

16. Dhiwahar, A.T., et al., Microwave-assisted combustion synthesis of pure and zinc-doped copper ferrite nanoparticles: Structural, morphological, optical, vibrational, and magnetic behavior. *Journal of Physics and Chemistry of Solids*, 2020. **138**: p. 109257.
17. Zeng, X., et al., Direct observation of cation distributions of ideal inverse spinel $CoFe_2O_4$ nanofibres and correlated magnetic properties. *Nanoscale*, 2017. **9**(22): p. 7493–7500.
18. Lininger, C.N., et al., Energetics of lithium insertion into magnetite, defective magnetite, and maghemite. *Chemistry of Materials*, 2018. **30**(21): p. 7922–7937.
19. Jia, J., et al., New insights into the magnetic characteristics of high mountain loess in Central Asia and its paleoclimatic implications. *Quaternary International*, 2019. **502**: p. 71–77.
20. Al-Khattawi, A. and A.R. Mohammed, Challenges and emerging solutions in the development of compressed orally disintegrating tablets. *Expert Opinion on Drug Discovery*, 2014. **9**(10): p. 1109–1120.
21. Thakare, Y., et al., Nanofluid-based drug delivery systems, in *Applications of nanofluids in chemical and bio-medical process industry*. 2022, Elsevier. p. 303–334.
22. Navarrete, C., et al., Cell factories for industrial production processes: Current issues and emerging solutions. *Processes*, 2020. **8**(7): p. 768.
23. Chandrakala, V., V. Aruna, and G. Angajala, Review on metal nanoparticles as nanocarriers: Current challenges and perspectives in drug delivery systems. *Emergent Materials*, 2022. **5**(6): p. 1593–1615.
24. Tiwari, S., et al., Comprehensive in-vitro and magnetic hyperthermia investigation of biocompatible non-stoichiometric $Zn_{0.5}Ca_{0.5}Fe_2O_4$ and $Mg_{0.5}Ca_{0.5}Fe_2O_4$ nanoferrites on lung cancer cell lines. *Journal of Alloys and Compounds*, 2024. **972**: p. 172588.
25. El Heda, I., et al., Structural, morphological, optical, magnetic and impedance spectroscopic properties of $Ni_{0.5}Fe_2\ 5O_4$ nanoparticles synthesized by the self-combustion method. *Physica E: Low-dimensional Systems and Nanostructures*, 2024. **158**: p. 115904.
26. Dang, Y. and J. Guan, Nanoparticle-based drug delivery systems for cancer therapy. *Smart Materials in Medicine*, 2020. **1**: p. 10–19.
27. Rao, D., et al., Role of nanoparticles in drug delivery. *International Journal of Nanotechnology and Applications*, 2010. **4**(1): p. 45–49.
28. Adepu, S. and S. Ramakrishna, Controlled drug delivery systems: Current status and future directions. *Molecules*, 2021. **26**(19): p. 5905.
29. Kalaiselvan, C.R., et al., Manganese ferrite ($MnFe_2O_4$) nanostructures for cancer theranostics. *Coordination Chemistry Reviews*, 2022. **473**: p. 214809.
30. Dahiya, M.S., V.K. Tomer, and S. Duhan, Metal–ferrite nanocomposites for targeted drug delivery, in *Applications of Nanocomposite Materials In Drug Delivery*. 2018, Elsevier. p. 737–760.
31. Salunkhe, A.B., M.V. Londhe, and V.M. Khot, Ferrite-and non-ferrite-based superparamagnetic materials, in *Superparamagnetic Materials for Cancer Medicine*. 2023, Springer. p. 57–72.
32. Pusta, A., et al., Recent advances in the development of drug delivery applications of magnetic nanomaterials. *Pharmaceutics*, 2023. **15**(7): p. 1872.
33. Gautam, S., V. Thakur, and N. Goyal, Nanoferrites as drug carriers in targeted drug delivery applications, in *Applications of Nanostructured Ferrites*. 2023, Elsevier. p. 161–178.
34. Sahoo, P., et al., Recent advances in zinc ferrite ($ZnFe_2O_4$) based nanostructures for magnetic hyperthermia applications. *Chemical Communications*, 2023. **59**(81): p. 12065–12090.

35. Vinod, S. and J. Philip, Thermal and rheological properties of magnetic nanofluids: Recent advances and future directions. *Advances in Colloid and Interface Science*, 2022. **307**: p. 102729.
36. Mokhosi, S.R., et al., Advances in the synthesis and application of magnetic ferrite nanoparticles for cancer therapy. *Pharmaceutics*, 2022. **14**(5): p. 937.
37. Puri, R. and V. Arora. Magnetic nano-ferrites as targeted drug delivery system for the treatment of cancer. in *AIP Conference Proceedings*. 2023, AIP Publishing.
38. Ghazimoradi, M., et al., Core-triple shells pH-responsive zinc-ferrite nanocomposite: Efficient dual remedy for targeted lung cancer therapy. *Journal of Drug Delivery Science and Technology*, 2023. **88**: p. 104950.
39. Singh, S., et al., Enhanced therapeutic action of Trastuzumab loaded $Zn_xMn_{1-x}Fe_2O_4$ nanoparticles using a pre-treatment step for hyperthermia treatment of HER^{2+} breast cancer. *Colloids and Surfaces B: Biointerfaces*, 2023. **232**: p. 113579.
40. Mohana, S. and S. Sumathi, Agaricus bisporus mediated synthesis of cobalt ferrite, copper ferrite and zinc ferrite nanoparticles for hyperthermia treatment and drug delivery. *Journal of Cluster Science*, 2023. **5**(1): p. 129–142.
41. Tsave, O., et al., Functional polysaccharide-coated SPIONs for in vitro mRNA delivery in breast cancer cells. *Materials Advances*, (2024). **5**(13): p. 5410–5432.
42. Mohammadpour-Haratbar, A., Y. Zare, and K.Y. Rhee, A review on drug delivery systems containing polymer nanocomposites for breast cancer treatment. *Polymer Reviews*, 2023. **64**(2): p. 490–527.
43. Sharma, A., R.C. Thakur, and R. Sharma, Biomedical applications of ferrites, in *Engineered ferrites and their applications*. 2023, Springer. p. 241–256.
44. Gadade, D.D., et al., Strategies for cancer targeting: Novel drug delivery systems opportunities and future challenges, in *Targeted cancer therapy in biomedical engineering*. 2023, Springer. p. 1–42.
45. Falsafi, S.R., et al., Metal nanoparticles and carbohydrate polymers team up to improve biomedical outcomes. *Biomedicine & Pharmacotherapy*, 2023. **168**: p. 115695.
46. Ameen, F. and N. Majrashi, Recent trends in the use of cobalt ferrite nanoparticles as an antimicrobial agent for disability infections: A review. *Inorganic Chemistry Communications*, 2023. **156**: p. 111187.
47. Tang, P., et al., Challenges and opportunities for improving the druggability of natural product: Why need drug delivery system? *Biomedicine & Pharmacotherapy*, 2023. **164**: p. 114955.
48. Bordone, M. and A. Bettencourt, Management of bone diseases: Looking at scaffold-based strategies for drug delivery. *Drug Delivery and Translational Research*, 2023. **13**(1): p. 79–104.
49. Pantwalawalkar, J., et al., Stimuli-responsive design of metal–organic frameworks for cancer theranostics: Current challenges and future perspective. *ACS Biomaterials Science & Engineering*, 2023. **9**(8): p. 4497–4526.
50. Mustafai, A., et al., Recent progress in proteins-based micelles as drug delivery carriers. *Polymers*, 2023. **15**(4): p. 836.

6 Antimicrobial Activities of Spinel and Inverse Spinel Ferrites

Shubham Kumar Patial, Anuj Sharma, Neelam Vishwakarma and Suman Singh

6.1 INTRODUCTION

In recent years, the extensive overuse of antibiotics has led to the widespread proliferation of antibiotic-resistant bacteria, resulting in an urgent, serious threat to public health on a global scale [1, 2]. This resistance development is primarily attributed to the *NDM-1* gene, encoding the New Delhi metallo-beta-lactamase-1 enzyme, which hydrolyzes and renders inactive beta-lactam antibiotics, including carbapenems, which are often regarded as the last line of defense against multidrug-resistant bacteria [3–5]. The surge in antibiotic resistance has given rise to so-called "superbugs," making them increasingly challenging and, in some instances, impossible to treat using conventional antibiotic therapies [6].

This escalating concern has spurred researchers to delve into alternative antimicrobial agents, exploring various materials with biomedical applications. Nanoparticles, in particular, have attracted much attention for their potential to assist drug molecules [7]. Moreover, many nanomaterials are being investigated for their antimicrobial properties, offering a promising approach to combat the growing threat of antimicrobial resistance. Among these materials, ferrites have emerged as particularly promising candidates due to their intrinsic antimicrobial properties, attributed to their capacity to generate reactive oxygen species (ROS) under specific conditions. ROS, such as hydroxyl radicals ($OH^{\cdot}$) and superoxide ions ($O_2^{\cdot-}$), exhibit potent antimicrobial activity by damaging bacterial cell membranes, proteins, and DNA, thereby inhibiting microbial growth and proliferation [8–10].

Ferrites, classified as a class of ceramic compounds, are composed of iron oxide (Fe_2O_3) combined with various metallic elements, such as cobalt, nickel, zinc, or manganese [11]. These materials possess a crystalline structure known as spinel, characterized by a face-centered cubic (FCC) lattice with oxygen ions at the corners and center of each face, while metal ions occupy the interstitial sites (Figure 6.1) [12, 13]. This unique structure also imparts exceptional magnetic properties to ferrites, including high saturation magnetization, coercivity, and magnetic anisotropy, making them vital for a wide variety of technological applications, including

DOI: 10.1201/9781032665658-6

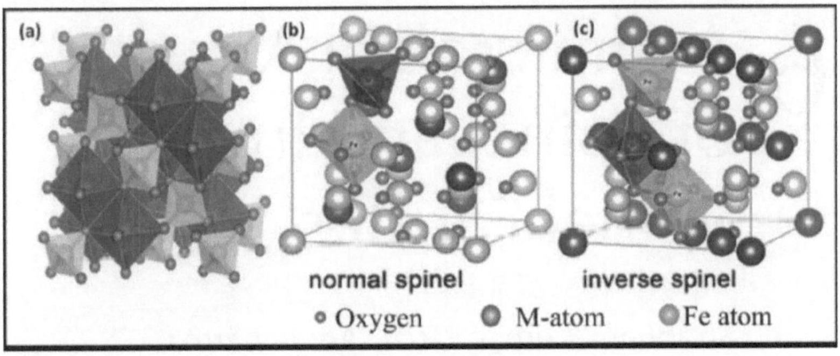

FIGURE 6.1 Structure of magnetic spinel ferrite showing tetrahedral sites (yellow), octahedral sites (green), and oxygen atoms (red) units (a) [15]. The unit cell structure of (b) standard spinel ferrite and (c) inverse spinel ferrite [16]. It is reprinted with permission.

magnetic storage media, sensors, magnetic separation, medical imaging, hyperthermia therapy, and targeted drug delivery applications [11]. Spinel ferrites, such as magnetite (Fe_3O_4) and maghemite (γ-Fe_2O_3), exhibit robust magnetic properties and biocompatibility, making them attractive candidates for biomedical applications [14]. Additionally, researchers are exploring various inverse spinel ferrites, which possess altered cation distributions within the spinel structure, leading to modifications in their magnetic and electronic properties.

Studies have demonstrated that the antimicrobial activity of spinel and inverse spinel ferrites can be enhanced through various approaches, including surface functionalization, size and morphology optimization, and doping with transition metal ions [13, 17]. Nigam et al. used single-phase spinel ferrites with spiked zinc doping and observed a proportional increase in magnetic properties with rising zinc concentrations [18]. The study successfully loaded 36.6% gentamicin drug for antimicrobial properties and had enhanced loading and release capabilities against *E. coli* bacteria. Iqbal et al. evaluated the antibacterial and antitumor response of $Zn_{0.5}Co_{0.5}Fe_2O_4$ nanoparticles (NPs) synthesized using the sol–gel method with a 60% loss in 250 µg/mL nanoparticle solution in cell viability examination [19]. Maksoud et al. examined the synthesis of $M_xCo_{(1-x)}Fe_2O_4$ with zinc, copper, and manganese and observed significant antimicrobial properties against multidrug-resistant pathogens [20]. These strategies enable researchers to refine and modify ferrite nanoparticles' physical and chemical properties, thereby improving their antimicrobial efficacy and biocompatibility for specific biomedical and environmental applications.

6.2 SYNTHESIS METHODS OF SPINEL AND INVERSE SPINEL FERRITES

Various synthesis methods and conditions play crucial roles in determining the quality and properties of spinel and inverse spinel ferrite nanoparticles (SISFNs). These nanoparticles are essential in numerous applications, including antimicrobial activities. Several synthesis techniques have been employed for SISFNs, encompassing co-precipitation [21], vapor deposition [22], thermal decomposition [23], hydrothermal, solvothermal [24], sol–gel [25], sonochemical [26], microwave-assisted [27], electrochemical [28], and mechanical milling techniques [29].

Among these methods, sonochemical method and microwave irradiation have gained prominence due to their advantages, such as milder reaction conditions, higher yields, improved selectivity, and cleaner reactions [30]. Each synthesis method has its merits and demerits, with parameters like temperature, solvent type, pH, purity of reactants, and reaction time significantly influencing the quality of synthesized SISFNs. For instance, the crystallite size of SFNs is known to depend on factors such as reaction time, pH, and temperature [31].

SISFNs are commonly synthesized for antimicrobial applications using co-precipitation or thermal decomposition methods. Co-precipitation synthesis requires careful adjustment of parameters such as pH, ionic strength, temperature, and the mole ratio of divalent metal to trivalent ion (Fe^{3+}). This method offers simplicity and environmental friendliness, with the size of SISFNs controlled by temperature or stirring time. Thermal decomposition, on the other hand, allows for precise control over the size and shape of SISFNs by adjusting temperature, heating rate, precursor concentration, and solvent amount [32,33].

Thermal decomposition is preferred over co-precipitation for antimicrobial purposes because it produces SISFNs with high crystallinity and narrow size distributions. In contrast, co-precipitation synthesis often results in SISFNs with a broad size distribution and a tendency to aggregate. Both methods, however, offer scalability for industrial and medical applications [34,35].

Surface coating with biocompatible materials is feasible during thermal decomposition synthesis, further enhancing SISFN properties. Additionally, the "heat-up" method has been described as suitable for the scalable synthesis of uniform iron oxide NPs, particularly for medical applications. This method involves slow heating of a reaction mixture composed of precursors, surfactants, and high boiling solvents [36,37].

6.3 SPINEL AND INVERSE SPINEL FERRITES FOR ANTIMICROBIAL APPLICATIONS

This section concisely overviews several spinel and inverse spinel ferrites, emphasizing their relevance in antimicrobial applications. The selection of $CoFe_2O_4$, Fe_3O_4, γ-Fe_2O_3, $MnFe_2O_4$, $NiFe_2O_4$, and $ZnFe_2O_4$ nanoparticles (NPs) is based on their extensive exploration in the biomedical field. $CoFe_2O_4$, recognized for its magnetic properties and high anisotropy, is a promising candidate for diverse medical applications, including drug delivery, hyperthermia-based tumor treatment, antimicrobial applications, and magnetic resonance imaging (MRI). Notably, due to their slower magnetic moment relaxation, $CoFe_2O_4$ NPs exhibit superior performance in antimicrobial activities compared to Fe_3O_4 and γ-Fe_2O_3 NPs. Doping $CoFe_2O_4$ with trace amounts of Zn^{2+} or other metals enhances its magnetic properties and anticancer activities [38]. Fe_3O_4, a widely studied soft magnet, finds extensive use in MRI, magnetically guided drug delivery, and hyperthermia-based cancer therapy. Similarly, γ-Fe_2O_3 NPs, known for their biocompatibility and superparamagnetic properties, are utilized in MRI contrast enhancement, biomagnetic separation, and magnetic drug delivery systems. $MnFe_2O_4$, characterized by its soft magnetic nature and high magnetic permeability, is employed in biomedical applications such as MRI contrast enhancement and hyperthermia-based cancer treatment. $NiFe_2O_4$ NPs, recognized for their low cost and moderate magnetic properties, serve as core materials in transformers, inductors, and biomedical devices, offering non-cytotoxicity, antimicrobial properties, and potential for cancer treatment by hyperthermia. Zinc ferrite ($ZnFe_2O_4$) NPs, exhibiting ferromagnetic characteristics in nanoparticulate form, demonstrate promise in cancer treatment by hyperthermia, mainly when synthesized via methods yielding smaller particle sizes and higher magnetization values.

6.3.1 Antimicrobial Testing Methods

Antimicrobial susceptibility tests (ASTs) are methods that can provide the choice of the best therapeutic option for microbial infection treatment [39]. In other words, if a pathogen's antimicrobial resistance (AMR) rate is greater than 20%, the drug should not be administered solely for treatment [40]. Clinical laboratories employ several of these ASTs to characterize drugs and metabolites successfully. This section covers the different tests that have been used both traditionally and in the modern day, the evaluation criteria, and the interpretation of the results.

6.3.2 Standard Testing Protocols

Conventional ASTs are based on the response of microbes in the presence of an antimicrobial agent. Traditional methods of ASTs were culture-dependent and included the dilution method (broth and agar), antimicrobial gradient method, disk diffusion test, chromogenic agar media, and colorimetric tests as shown in Figure 6.2 [39, 40]. These methods were firmly established in the diagnostic routine, but their main drawback was the lag phase in the growth of bacteria, which proved to be 18–24 h in time (prolonged up to 48 h for some species) [41]. Recent progress and development

Traditional methods	Automated and Semi-Automated methods	MS
• Dilution method • Antimicrobial gradient method • Disk diffusion test • Chromogenic agar media • Colorimetric tests	• PCR • DNA microarrays • Whole genome sequencing	• MALDI-TOF MS

FIGURE 6.2 Different AST procedures for antimicrobial activity testing.

in AST methodology have produced new and improved techniques that overcome the drawbacks of traditional methods. These technologies include automated and semi-automated devices based on microdilution susceptibility testing, molecular-based techniques (polymerase chain reaction, DNA microarrays, and whole-genome sequencing), and mass spectrometry (MS) [40, 41]. Yoon et al. demonstrated in their study that MALDI-TOF MS is the most suitable for AMR strains like MRSA, VRE, CRAB, and CRPA due to its speed and simple application [42].

6.3.3 Evaluation Criteria

Since many antimicrobial agents can be used because of their antimicrobial activity, it would be strenuous to test each and every agent for its antimicrobial efficacy. For this, a laboratory can individually decide and test various seemingly fit agents, following some common rules and regulations [40]. The evaluation of the agent is done based on the area cleared by the agent, microbial susceptibility once the agent is removed, the organism isolated, and the agent's toxicity. The study performed is primarily focused on producing antimicrobial agents that are non-toxic to animals and plants but can show antimicrobial properties [43].

6.3.4 Interpretation of Results

Interpreting results from ASTs and reporting the susceptibility categories to antimicrobial agents require breakpoints from the two most commonly used systems: CLSI (Clinical and Laboratory Standards Institute) and EUCAST (European Committee on Antimicrobial Susceptibility Testing). CLSI breaks down recommendations for important agents that should be tested routinely (group A) and those that can be tested without routine as they don't pose a threat (groups B and C) [39–41]. In the upcoming sections, the antimicrobial properties of spinel and inverse spinel ferrites are discussed.

6.4 ANTIMICROBIAL ACTIVITIES SPINEL FERRITES AND INVERSE SPINEL FERRITES

Bacterial infections have become a pressing global health concern due to the proliferation of antibiotic-resistant bacterial strains, necessitating the development of novel antibacterial agents. In this regard, inorganic antimicrobial agents have gained prominence over their organic counterparts due to their superior heat resistance and longevity [44]. In this context, inorganic agents are more potent and easier to use. Among the various candidates explored, nanoferrite materials have garnered attention for their promising antibacterial properties. In recent years, there has been a surge in research on spinel and inverse spinel ferrites and their application as antimicrobial agents in waste treatment [45].

The antimicrobial activity of any material describes its microbicidal property. Several materials have been tested, and their antimicrobial activity has been recorded [40]. Ferrite NPs are among the materials that show significant antimicrobial activity, which can be increased by introducing metal ions into the ferrite core. The hybridization of metal cores and ferrite ions determines the configuration of spinel and inverse spinel ferrites. These materials are considered because of their synthesis procedures (green synthesis routes are available), their antimicrobial activity (substituted metal ions in the ferrite core), and their nanostructures (easy to penetrate the cell wall) [41, 44, 46].

Notably, ferrite nanoparticles (FNPs) have demonstrated efficacy against Gram-negative and Gram-positive bacteria. For instance, Samavati and Ismail synthesized copper-substituted cobalt FNPs, which exhibited enhanced antibacterial activity against *Escherichia coli* (*E. coli*) due to the presence of copper ions [47]. Similarly, Maksoud et al. studied metal-substituted cobalt FNPs. They found that zinc-doped cobalt ferrites showed significant inhibition against various harmful bacterial strains, including *Staphylococcus aureus* (*S. aureus*) and *Enterococcus columbae* [48]. Hathout et al. demonstrated the use of $CoFe_2O_4$ (inverse spinel ferrite of cobalt) NPs as an antimicrobial agent for medical applications [49]. In this section, the antimicrobial activity of spinel ferrites and inverse spinel ferrites and their mechanisms are discussed.

Moreover, the antibacterial potential of FNPs extends to biogenic synthesis methods. Naik et al. utilized *Limonia acidissima* juice to synthesize biogenic zinc ferrite nanoparticles, which displayed notable inhibition against *S. aureus* and *E. coli.* The mechanism behind their antibacterial effect involves the production of reactive oxygen species (ROS) upon light exposure, leading to bacterial cell damage. Additionally, zinc-doped cobalt ferrite nanoparticles synthesized using curd as a fuel exhibited potent inhibitory effects against both Gram-positive and Gram-negative bacteria, attributed to ROS generation and the release of metal ions. These findings highlight the diverse strategies employed in synthesizing FNPs for antimicrobial applications, ranging from chemical to biogenic approaches [50].

Furthermore, the antibacterial efficacy of FNPs can be enhanced through compositional modifications. Studies by Almessiere et al. demonstrated the synthesis of various compositions of manganese-zinc ferrites and their superior bactericidal efficacy against both Gram-positive and Gram-negative bacteria [51]. Similarly, Abdel Maksoud et al. reported that magnesium-substituted manganese-zinc ferrite nanoparticles exhibited significant inhibitory action against a wide range of pathogenic bacteria due to their narrow band gap, which facilitates the generation of ROS upon light exposure [52]. These examples underscore the importance of compositional tuning in optimizing the antimicrobial properties of FNPs.

Figure 6.3 illustrates the antibacterial properties of $Ni_{0.5}Zn_{0.5}Fe_2O_4$ nanoparticles (NPs). The mechanism primarily involves the disruption of bacterial cell walls by metal oxides. These nanoparticles generate various reactive oxygen species, including hydroxyl radicals ($OH^{\cdot}$), the less harmful superoxide anion (O^{2-}), and hydrogen peroxide (H_2O_2), a weak oxidizer. This activity is largely due to factors such as the polar nature of the NP surface, a high specific surface area, distinct morphology, grain size, and an increase in oxygen vacancies. These features facilitate the diffusion of chemical molecules and the release of Ni^{2+}, Zn^{2+}, and Fe^{3+} ions [53, 54].

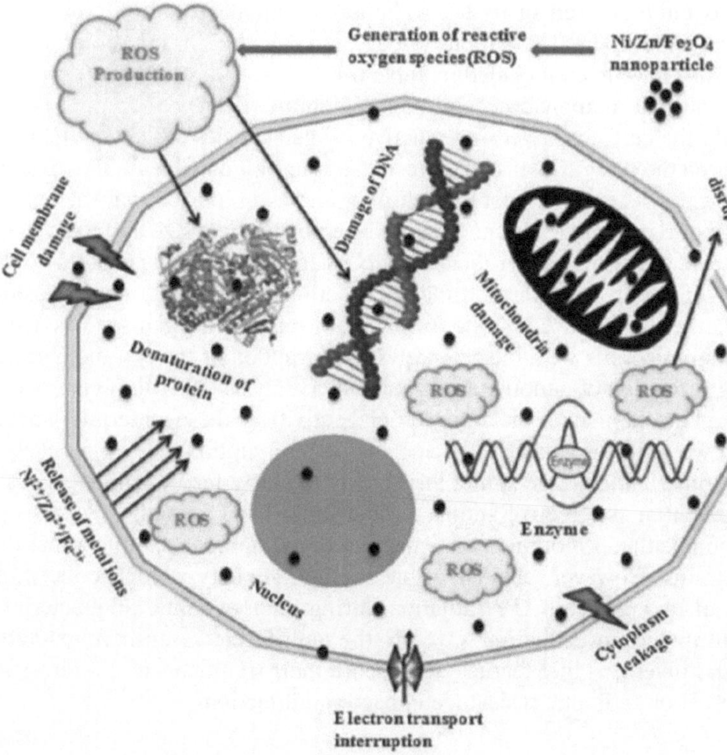

FIGURE 6.3 Potential mechanisms underlying the antibacterial efficacy of Ni/Zn ferrite nanoparticles [57].

The interaction between the positively charged Ni^{2+}, Zn^{2+}, and Fe^{3+} ions and the negatively charged bacterial cell membranes leads to their attraction. This interaction disrupts DNA replication and protein structures within the bacteria, ultimately leading to cell death [55, 56]. Specifically, the Ni/Zn ferrite nanoparticles demonstrate substantial effectiveness against *Pseudomonas aeruginosa*.

The smaller crystalline size of these nanoparticles is particularly noted for enhancing their antibacterial effectiveness. The results indicate that Ni/Zn/Fe_2O_4 NPs are highly effective against human pathogens and are promising candidates for use in pharmaceutical applications [58, 59].

The use of inverse spinel ferrites in their nanocomposite form has been reported in several studies. Ma et al. demonstrated the successful deposition of Ag and $CoFe_2O_4$ nanoparticles on the GO surface to form an Ag-$CoFe_2O_4$-GO composite that was used to inactivate bacteria and remove Pb(II) from wastewater photocatalytically. They reported the inactivation of *E. coli* and *S. aureus* from wastewater [60]. Another study by Rajabi et al. worked on the mode of action and the relation of the particle size with its antimicrobial activity. They reported that a decreased particle size and increased surface-to-volume ratio increased antimicrobial activity [61]. Several experimentally demonstrated studies of the antimicrobial activity of these materials have been included in Table 6.1.

These materials demonstrate antimicrobial properties by incorporating themselves inside the cells and releasing reactive oxygen species (ROS) that have apoptotic effects on the cell [45]. Furthermore, these materials can release the toxic metals they store in their structure's core. Hashim et al. have reported that the antimicrobial activity of these materials can be influenced by particle size and that nanosized particles have greater antimicrobial activity than micro-sized particles [69]. The mechanism of action for their antimicrobial activity includes (i) decomposition of the ferrite structure followed by the formation of reactive oxygen species, (ii) interaction of nanomaterials with the cell wall and disruption of the cell membrane, and (iii) optical activation of nanomaterials via photocatalysis [39, 40]. Figure 6.4 shows a generalized mechanism of the antimicrobial activity of these materials. They work in different ways, like releasing ROS and cell wall disruption.

Hence, spinel and inverse spinel ferrites, particularly ferrite nanoparticles, hold immense potential as effective antibacterial agents. Their versatile synthesis methods and compositional modifications enable tailored approaches to combat diverse bacterial strains. However, challenges such as cytotoxicity at high concentrations and potential toxicity from UV radiation during photodynamic antibacterial therapy warrant further investigation. Overall, the multifaceted antimicrobial activities of spinel and inverse spinel ferrites underscore their significance in addressing the growing threat of antibiotic resistance in bacterial infections.

6.5 FACTORS INFLUENCING ANTIMICROBIAL PROPERTIES

The antimicrobial properties of these materials can be attributed to the cation attraction of the microbial protein material, which further gets precipitated in the form of insoluble proteinate that leads to cell death [70]. Inverse spinel ferrites are core–shell nanoparticles with tunable properties and high functionality. The metal ions in the core of the structure enhance the antimicrobial activity [71]. This section covers their composition, properties, and effects on the environment.

Inverse spinel ferrites have a general chemical formula of ($[M^{2+}\ Fe^{3+}]B\ [Fe^{3+}]AO_4$) where the M^{2+} is the cation, and that occupies octahedral sites B and Fe^{3+} occupies both tetrahedral sites A and octahedral sites B (e.g., $FeOFe_2O_3$, $NiFe_2O_4$, and $CdFe_2O_4$) [39, 41, 44, 45, 70]. These materials have various advantages, such as good catalytic activity, easy synthesis routes, availability of green synthesis routes, simplicity of operation, and environment-friendliness [46, 71]. These advantages, combined with their magnetic properties, make them good candidates for reusability.

Particle size and morphology of these materials are crucial components of their antimicrobial activity. Several reports have demonstrated the direct relation between particle size and microbicidal properties [39]. Alsafari et al. concluded that the inactivation of infectious strains by inverse spinel ferrites could be due to the penetrability of nanoparticles inside the cells, which could cause oxidative stress that results in cell death [72]. The properties of these materials are not only related to their size

TABLE 6.1
Spinel and Inverse Spinel Ferrites Were Used, and Their Core Metal and AST Were Used to Determine Their Antimicrobial Activity

Ferrite Composition	Synthesis Method	Core Metal Ion	Antimicrobial Susceptibility Test Used	References
$[(CuCr_xFe_{2-x}O_4)(0.0 < x < 1.0)]$	Co-precipitation	Cr, Cu	Disk diffusion test	[62]
Cu-$MgFe_2O_4$	Co-precipitation	Cu, Mg	Agar well diffusion test	[63]
$[(Cr_xNi_{1-x}Fe_2O_4)(0.0 < x < 0.210)]$	Sol–gel auto-combustion (SGCAT) method	Cr, Ni	Disk diffusion test	[64]
$[(Zn_xCo_{1-x}Fe_2O_4)$(x in range of 0.08–0.56)	Sol–gel auto-combustion (SGCAT) method	Co, Zn	Colorimetric test	[65]
$CuFe_2O_4$	Sol–gel auto-combustion (SGCAT) method	Cu	Broth microdilution assay	[66]
$ZnFe_2O_4$	Sol–gel auto-combustion (SGCAT) method	Zn	Broth microdilution assay	[66]
$MnFe_2O_4$	Sol–gel auto-combustion (SGCAT) method	Mn	Broth microdilution assay	[66]
$ZnFe_2O_4$	Microwave-assisted solution combustion method	Zn	Agar dispersion and agar well diffusion methods	[67]
$CuFe_2O_4$	Co-precipitation technique	Cu	Agar disk diffusion	[68]
$ZnFe_2O_4$	Co-precipitation technique	Zn	Agar disk diffusion	[68]
$[(Mn_{0.5}Zn_{0.5-x}Mg_xFe_2O_4)$ (x=0, 0.125, 0.25, 0.375, and 0.5)]	Sol–gel method	Mn, Zn, Mg	Agar dispersion and agar well diffusion methods	[52]

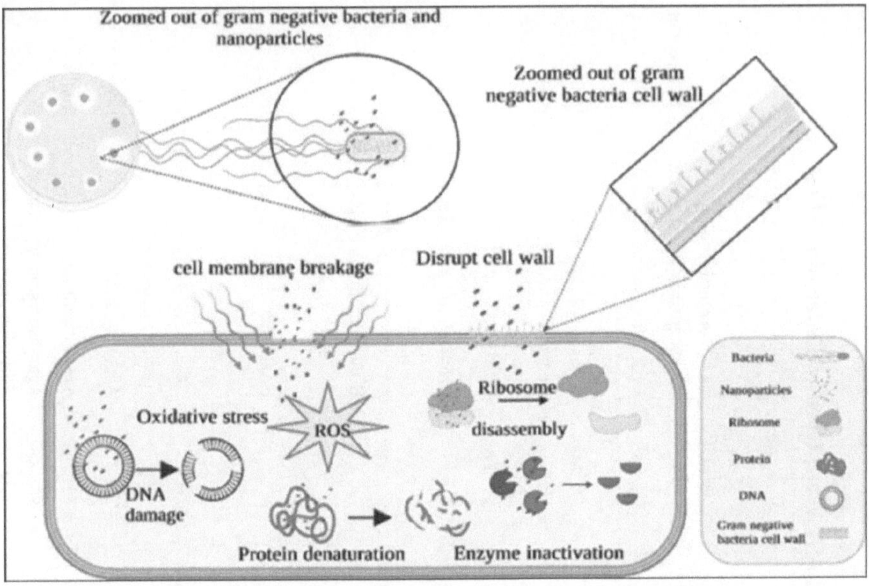

FIGURE 6.4 Schematic of action of nanoparticles in their antimicrobial activity [66].

but also to many other factors like the radii of cations, ionic sites present, Madelung energy, and field effects of the crystal [41]. The morphology governs the application, as a specific morphology of these materials can be better for certain applications.

These materials have a metal ion core and ferrite base with a general chemical formula of ($[M^{2+}Fe^{3+}]B\ [Fe^{3+}]AO_4$) where the M^{2+} is the cation, and that occupies octahedral sites B and Fe^{3+} occupies both tetrahedral sites A and octahedral sites B (e.g., $FeOFe_2O_3$, $NiFe_2O_4$, and $CdFe_2O_4$) [43, 71]. These materials have properties ranging from electrical conductivity to magnetic properties and additional antimicrobial properties. Surface functionalization includes the adsorption of reactive species that can help these materials in catalysis. The core divalent metal atom also governs the surface functionalization as certain metal ions (like Cu, Ni, etc.), when placed at the core of these structures, enhance antimicrobial activity [39, 40, 44].

As discussed earlier, in inverse spinel ferrites, Fe^{3+} occupies both tetrahedral and octahedral sites of the resulting structure. This creates a few empty octahedral sites that are filled by the divalent atom [40]. The catalytic activity of ferrites depends on various factors, like the nature of the divalent metal atoms, the synthesis method used, pollutant/microbial concentration, size of the resulting structure, pH, temperature, and light source for photo-Fenton-like oxidation [45, 70]. The stability and durability of these structures prohibit these factors from deviating from the catalytic activity, but other properties like adsorption and antimicrobial activity can be affected significantly [46, 70, 71]. The next section deals with the application and future scope of these materials.

6.6 POTENTIAL APPLICATIONS AND FUTURE PERSPECTIVES

The antimicrobial properties of spinel and inverse spinel ferrites present many potential applications in diverse fields, where their ability to inhibit the growth of harmful bacteria and fungi is highly valued. In the medical field, these ferrites can be employed in developing antimicrobial coatings for surgical instruments, implants, and hospital surfaces, thus reducing the risk of hospital-acquired infections and improving patient outcomes. Moreover, in the realm of water purification, spinel and inverse spinel ferrites could be utilized in filters and membranes to remove microbial contaminants, ensuring safe drinking water in both urban and remote areas. Another promising application lies in the food industry, where these materials can be integrated into packaging to prolong the shelf life of perishable goods by preventing microbial spoilage. In Figure 6.5, the characteristics and benefits of spinel and inverse spinel ferrites are depicted. These advantageous traits make them highly desirable for a wide array of applications, spanning from catalysis and sensing to biomedical uses.

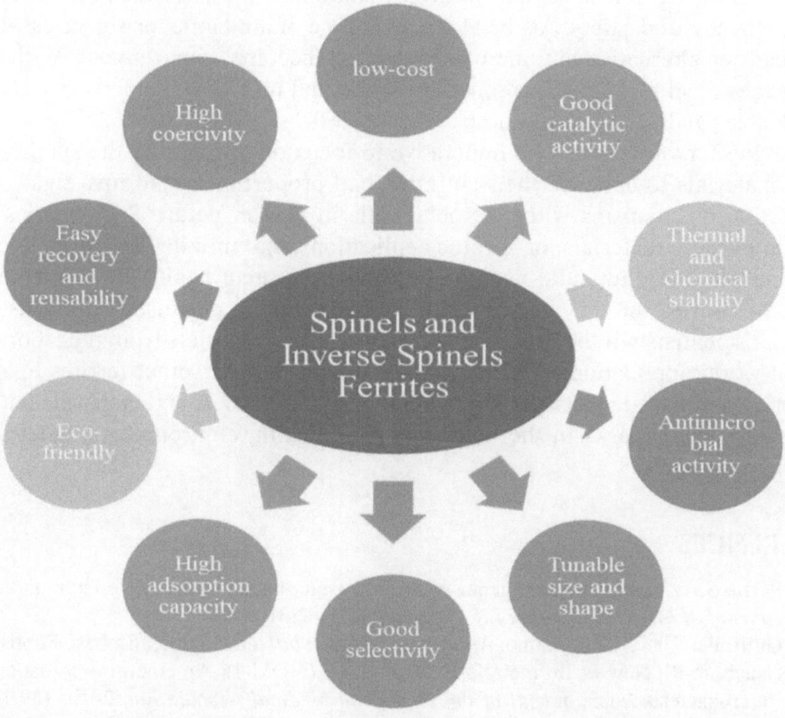

FIGURE 6.5 Properties and advantages of spinel and inverse spinel ferrites. These advantages promote using these materials for diverse applications ranging from catalysis to sensing to biomedical applications.

Looking to the future, the expansion of research into the nanoengineering of spinel and inverse spinel ferrites could unlock further applications. As synthetic antimicrobials' environmental and health implications become more pressing, ferrites offer a sustainable alternative due to their effectiveness, durability, and potential for safe integration into various matrices. Advances in synthesis methods will also be crucial, aiming to enhance antimicrobial efficacy while reducing costs and environmental impact. Furthermore, understanding the mechanisms by which these ferrites exert their antimicrobial effects could lead to designing tailored materials targeted at specific pathogens, enhancing the specificity and effectiveness of antimicrobial interventions. This convergence of material science, microbiology, and engineering heralds a new era of innovation in antimicrobial technologies.

6.7 CONCLUSION

In conclusion, the exploration of spinel and inverse spinel ferrites as antimicrobial agents opens a promising pathway for the development of novel, efficient, and environmentally friendly antimicrobial solutions. The research discussed in this chapter underscores the potential of these ferrites to combat a wide range of microbial pathogens, offering significant advantages over traditional antimicrobials in terms of stability, efficacy, and safety. As the global challenge of antibiotic resistance escalates, the need for alternative antimicrobial strategies becomes more urgent. With their unique chemical and physical properties, spinel and inverse spinel ferrites stand out as valuable candidates in this quest.

Moving forward, it will be imperative to focus on optimizing the synthesis of these materials to enhance their antimicrobial properties and to investigate their interaction mechanisms with microbial cells in greater detail. Such studies will help tailor these materials for specific applications, maximizing their effectiveness while minimizing potential negative impacts on human health and the environment. Moreover, interdisciplinary collaboration among chemists, biologists, and material scientists will be crucial in advancing these materials from the laboratory to real-world applications. Ultimately, spinel and inverse spinel ferrites hold the potential not only to meet the current demands for antimicrobial materials but also to set new benchmarks in the fields of public health, environmental safety, and beyond.

REFERENCES

1. Parra, S.A., et al., The emergence of drug resistant bacteria: Effects on human health. *Journal of Environment and Life Sciences*, 2017. **2**(3): p. 77–79.
2. Guilfoile, P. and I.E. Alcamo, *Antibiotic-resistant bacteria*. 2007, Infobase Publishing.
3. Hsueh, P.-R., New Delhi metallo-β-lactamase-1 (NDM-1): An emerging threat among Enterobacteriaceae. *Journal of the Formosan Medical Association*, 2010. **109**(10): p. 685–687.
4. Fattah, K., et al., Investigation of the efficacy of synthesized silver and zinc oxide nanoparticles against multi-drug resistant gram negative bacterial clinical isolates. *Archives of Clinical Microbiology*, 2017. **8**(6): p. 67.

5. Fazeli, H., et al., Detection of New Delhi metallo-beta-lactamase-1 (NDM-1) in carbapenem-resistant Klebsiella pneumoniae isolated from a university hospital in Iran. *Hippokratia*, 2015. **19**(3): p. 205.
6. Munguia, J. and V. Nizet, Pharmacological targeting of the host–pathogen interaction: Alternatives to classical antibiotics to combat drug-resistant superbugs. *Trends in Pharmacological Sciences*, 2017. **38**(5): p. 473–488.
7. Ravishankar Rai, V. and A. Jamuna Bai, Nanoparticles and their potential application as antimicrobials, in *Science against Microbial Pathogens: Communicating Current Research and Technological Advances*, Méndez-Vilas, A., Ed., 197–209. Spain: Formatex Research Center.
8. Ameen, F. and N. Majrashi, Recent trends in the use of cobalt ferrite nanoparticles as an antimicrobial agent for disability infections: A review. *Inorganic Chemistry Communications*, 2023. **156** p. 111187.
9. Ye, J., et al., Novel copper-containing ferrite nanoparticles exert lethality to MRSA by disrupting MRSA cell membrane permeability, depleting intracellular iron ions, and upregulating ROS levels. *Frontiers in Microbiology*, 2023. **14**: p. 1023036.
10. Rabbani, A., et al., Development of bactericidal spinel ferrite nanoparticles with effective biocompatibility for potential wound healing applications. *RSC Advances*, 2021. **11**(3): p. 1773–1782.
11. Tatarchuk, T., et al., Spinel ferrite nanoparticles: Synthesis, crystal structure, properties, and perspective applications, in *Nanophysics, nanomaterials, interface studies, and applications: Selected proceedings of the 4th international conference nanotechnology and nanomaterials (NANO2016), August 24–27, 2016, Lviv, Ukraine*. 2017, Springer.
12. Mounkachi, O., M. Hamedoun, and A. Benyoussef, Electronic and magnetic properties of $SnFe_2O_4$ spinel ferrites. *Journal of Superconductivity and Novel Magnetism*, 2017. **30**(11): p. 3035–3038.
13. Sanpo, N., et al., Multifunctional spinel ferrite nanoparticles for biomedical application. *Advanced Functional Materials*, 2015: p. 183–217.
14. Seehra, M., *Magnetic spinels: Synthesis, properties and applications*. 2017, BoD–Books on Demand.
15. Mmelesi, O.K., et al., Cobalt ferrite nanoparticles and nanocomposites: Photocatalytic, antimicrobial activity and toxicity in water treatment. *Materials Science in Semiconductor Processing*, 2021. **123**: p. 105523.
16. Narang, S.B. and K. Pubby, Nickel spinel ferrites: A review. *Journal of Magnetism and Magnetic Materials*, 2021. **519**: p. 167163.
17. Sayed, M.A., H. Abdelsalam, and A.A. El-Bassuony, Antimicrobial activity of Novel spinel nanoferrites against pathogenic fungi and bacteria. *World Journal of Microbiology and Biotechnology*, 2020. **36**(2): p. 25.
18. Nigam, A. and S. Pawar, Structural, magnetic, and antimicrobial properties of zinc doped magnesium ferrite for drug delivery applications. *Ceramics International*, 2020. **46**(4): p. 4058–4064.
19. Iqbal, S., et al., Structural, morphological, antimicrobial, and in vitro photodynamic therapeutic assessments of novel Zn^{+2}-substituted cobalt ferrite nanoparticles. *Results in Physics*, 2019. **15**: p. 102529.
20. Maksoud, M.A., et al., Synthesis and characterization of metals-substituted cobalt ferrite [$M_xCo_{(1-x)}Fe_2O_4$; (M = Zn, Cu and Mn; $x = 0$ and 0.5)] nanoparticles as antimicrobial agents and sensors for Anagrelide determination in biological samples. *Materials Science and Engineering: C*, 2018. **92**: p. 644–656.
21. Zi, Z., et al., Synthesis and magnetic properties of $CoFe_2O_4$ ferrite nanoparticles. *Journal of Magnetism and Magnetic Materials*, 2009. **321**(9): p. 1251–1255.

22. Zhang, D., et al., Low-temperature fabrication of $MnFe_2O_4$ octahedrons: Magnetic and electrochemical properties. *Chemical Physics Letters*, 2006. **426**(1): p. 120–123.
23. Ge, Q., et al., Hydrophilic superparamagnetic nanoparticles: Synthesis, characterization, and performance in forward osmosis processes. *Industrial & Engineering Chemistry Research*, 2011. **50**(1): p. 382–388.
24. Meidanchi, A., et al., $ZnFe_2O_4$ nanoparticles as radiosensitizers in radiotherapy of human prostate cancer cells. *Materials Science & Engineering C-Materials for Biological Applications*, 2015. **46**: p. 394–399.
25. Avazpour, L., et al., Synthesis of single-phase cobalt ferrite nanoparticles via a novel EDTA/EG precursor-based route and their magnetic properties. *Journal of Alloys and Compounds*, 2015. **637**: p. 497–503.
26. Zhang, S., et al., Sonochemical formation of iron oxide nanoparticles in ionic liquids for magnetic liquid marble. *Physical Chemistry Chemical Physics*, 2012. **14**(15): p. 5132–5138.
27. Mondal, A., et al., Microwave synthesis of a-Fe_2O_3 nanoparticles and their lithium storage properties: A comparative study. *Journal of Alloys and Compounds*, 2015. **648**: p. 732–739.
28. Fajaroh, F., et al., Synthesis of magnetite nanoparticles by surfactant-free electrochemical method in an aqueous system. *Advanced Powder Technology*, 2012. **23**: p. 328–333.
29. Zhang, Z., et al., Synthesis and characterization of nickel ferrite nanoparticles via planetary ball milling assisted solid-state reaction. *Ceramics International*, 2015. **41**(3, Part B): p. 4523–4530.
30. Safarik, I., et al., Microwave assisted synthesis of magnetically responsive composite materials. *IEEE Transactions on Magnetics*, 2013. **49**: p. 213–218.
31. Wang, F., et al., Hydrothermal synthesis and characterization of α-Fe_2O_3 nanoparticles. *Materials Science in Semiconductor Processing*, 2013. **16**: p. 802–806.
32. Humam, N., et al., Synthesis and characterization of $CaFe_2O_4$ nanoparticles via co-precipitation and auto-combustion methods. *Ceramics International*, 2017. **44**: p. 46–50.
33. Kefeni, K.K., T.A.M. Msagati, and B.B. Mamba, Ferrite nanoparticles: Synthesis, characterisation and applications in electronic device. *Materials Science and Engineering: B*, 2017. **215**: p. 37–55.
34. Laurent, S., et al., Magnetic iron oxide nanoparticles: Synthesis, stabilization, vectorization, physicochemical characterizations, and biological applications. *Chemical Reviews*, 2008. **108**(6): p. 2064–2110.
35. Monfared, A.H., et al., Reversible multistimuli-responsive manganese–zinc ferrite/P(NIPAAM-AAc-AAm) core-shell nanoparticles: A programmed ferrogel system. *Materials Chemistry and Physics*, 2019. **226**: p. 44–50.
36. Tartaj, P., et al., The preparation of magnetic nanoparticles for applications in biomedicine. *Journal of Physics D: Applied Physics*, 2003. **36**: p. R182–R197.
37. Lee, N., et al., Iron oxide based nanoparticles for multimodal imaging and magnetoresponsive therapy. *Chemical Reviews*, 2015. **115**(19): p. 10637–1089.
38. Amiri, S. and H. Shokrollahi, The role of cobalt ferrite magnetic nanoparticles in medical science. *Materials Science and Engineering: C*, 2013. **33**(1): p. 1–8.
39. CH, V., et al., Magnetic and antimicrobial properties of cobalt-zinc ferrite nanoparticles synthesized by citrate-gel method. *International Journal of Applied Ceramic Technology*, 2019. **16**(5): p. 1944–1953.
40. Bareke, H. and A.A. Oladipo, Antimicrobial activity of biogenic-synthesized novel bimetallic nanospinel LiCu-ferrite particles: Experimental and computational studies. *Journal of Molecular Structure*, 2024. **1296**: p. 136823.

41. Elayakumar, K., et al., Enhanced magnetic property and antibacterial biomedical activity of Ce^{3+} doped $CuFe_2O_4$ spinel nanoparticles synthesized by sol-gel method. *Journal of Magnetism and Magnetic Materials*, 2019. **478**: p. 140–147.
42. Yoon, E.-J. and S.H. Jeong, MALDI-TOF mass spectrometry technology as a tool for the rapid diagnosis of antimicrobial resistance in Bacteria. *Antibiotics*, 2021. **10**(8): p. 982.
43. Kumari, S., et al., Nano Ca–Mg–Zn ferrites as tuneable photocatalyst for UV light-induced degradation of rhodamine B dye and antimicrobial behavior for water purification. *Ceramics International*, 2023. **49**(8): p. 12469–12480.
44. Nigam, A. and S.J. Pawar, Structural, magnetic, and antimicrobial properties of zinc doped magnesium ferrite for drug delivery applications. *Ceramics International*, 2020. **46**(4): p. 4058–4064.
45. Fantozzi, E., et al., Silver doped magnesium ferrite nanoparticles: Physico-chemical characterization and antibacterial activity. *Materials*, 2021. **14**(11): p. 2859.
46. Almessiere, M.A., et al., Green synthesis of Nd substituted Co-Ni nanospinel ferrites: a structural, magnetic, and antibacterial/anticancer investigation. *Journal of Physics D: Applied Physics*, 2022. **55**(5): p. 055002.
47. Samavati, A. and A. F. Ismail, Antibacterial properties of copper-substituted cobalt ferrite nanoparticles synthesized by co-precipitation method. *Particuology*, 2017. **30**: p. 158–163.
48. Maksoud, M.I.A.A., et al., Antibacterial, antibiofilm, and photocatalytic activities of metals-substituted spinel cobalt ferrite nanoparticles. *Microbial Pathogenesis*, 2019. **127**: p. 144–158.
49. Amal S. Hathout, A.A., Bassem A. Sabry, Aziza A. El-Nekeety, Mohamed H. Roby, Nasseralla M. Deraz, Soher E. Aly, and Mosaad A. Abdel-Wahhab, Synthesis and characterization of cobalt ferrites nanoparticles with cytotoxic and antimicrobial properties. *Journal of Applied Pharmaceutical Science*, 2017. **7**(1): p. 86–92.
50. Naik, M.M., et al., Green synthesis of zinc doped cobalt ferrite nanoparticles: Structural, optical, photocatalytic and antibacterial studies. *Nano-Structures & Nano-Objects*, 2019. **19**: p. 100322.
51. Almessiere, M.A., et al., Synthesis of Dy-Y co-substituted manganese-zinc spinel nanoferrites induced anti-bacterial and anti-cancer activities: Comparison between sonochemical and sol-gel auto-combustion methods. *Materials Science and Engineering: C*, 2020. **116**: p. 111186.
52. Abdel Maksoud, M.I.A., et al., Influence of Mg^{2+} substitution on structural, optical, magnetic, and antimicrobial properties of Mn–Zn ferrite nanoparticles. *Journal of Materials Science: Materials in Electronics*, 2020. **31**(3): p. 2598–2616.
53. Lakshmanan, A., et al., Effect of fuel content on nonlinear optical and antibacterial activities of $Zn/Cu/Al_2O_4$ nanoparticles prepared by microwave-assisted combustion method. *Journal of King Saud University – Science*, 2020. **32**(2): p. 1382–1389.
54. Kannan, K., et al., Nanostructured metal oxides and its hybrids for photocatalytic and biomedical applications. *Advances in Colloid and Interface Science*, 2020. **281**: p. 102178.
55. Verma, R., et al., Ni addition induced modification of structural, magnetic properties and antistructural modeling of $Zn_{1-x}Ni_xFe_2O_4$ (x = 0.0 - 1.0) nanoferrites. *Molecular Crystals and Liquid Crystals*, 2018. **674**(1): p. 130–141.
56. Karthik, K., et al., Facile microwave-assisted green synthesis of NiO nanoparticles from Andrographis paniculata leaf extract and evaluation of their photocatalytic and anticancer activities. *Molecular Crystals and Liquid Crystals*, 2018. **673**(1): p. 70–80.

57. Lakshmanan, A., et al., Investigations on structural, optical, dielectric, electronic polarizability, Z-scan and antibacterial properties of Ni/Zn/Fe2O4 nanoparticles fabricated by microwave-assisted combustion method. *Journal of Photochemistry and Photobiology A: Chemistry*, 2020. **402**: p. 112794.
58. Aswini, R., S. Murugesan, and K. Kannan, Bio-engineered TiO2 nanoparticles using Ledebouria revoluta extract: Larvicidal, histopathological, antibacterial and anticancer activity. *International Journal of Environmental Analytical Chemistry*, 2021. **101**(15): p. 2926–2936.
59. Kannan, K., et al., Facile microwave-assisted synthesis of metal oxide CdO-CuO nanocomposite: Photocatalytic and antimicrobial enhancing properties. *Optik*, 2020. **218**: p. 165112.
60. Ma, S., et al., Highly efficient antibacterial and Pb(II) removal effects of Ag-$CoFe_2O_4$-GO nanocomposite. *ACS Applied Materials & Interfaces*, 2015. **7**(19): p. 10576–10586.
61. Rajabi, S.K. and S. Sohrabnezhad, Fabrication and characteristic of Fe_3O_4@MOR@CuO core-shell for investigation antibacterial properties. *Journal of Fluorine Chemistry*, 2018. **206**: p. 36–42.
62. Ansari, M.A., et al., Synthesis and characterization of antibacterial activity of spinel chromium-substituted copper ferrite nanoparticles for biomedical application. *Journal of Inorganic and Organometallic Polymers and Materials*, 2018. **28**(6): p. 2316–2327.
63. George, M., et al., Evaluation of Cu–MgFe2O4 spinel nanoparticles for photocatalytic and antimicrobial activates. *Journal of Physics and Chemistry of Solids*, 2021. **153**: p. 110010.
64. Bhasin, V., et al., Effect of Cr substitution on nickel spinel ferrite's surface morphology, structure, antibacterial activity and magnetic properties. *Inorganic Chemistry Communications*, 2024. **160**: p. 111764.
65. Omelyanchik, A., et al., Green synthesis of Co-Zn spinel ferrite nanoparticles: Magnetic and intrinsic antimicrobial properties. *Materials*, 2020. **13**(21): p. 5014.
66. Dabagh, S., S.A. Haris, and Y.N. Ertas, Synthesis, characterization and potent antibacterial activity of metal-substituted spinel ferrite nanoparticles. *Journal of Cluster Science*, 2023. **34**(4): p. 2067–2078.
67. Ansari, M.R., et al., Structural, optical, magnetic and anti-bacterial properties of green synthesized spinel zinc ferrite by microwave-assisted method. *Materials Chemistry and Physics*, 2023. **301**: p. 127641.
68. Hatami Kahkesh, K., et al., Synthesis, characterization, antioxidant and antibacterial activities of zinc ferrite and copper ferrite nanoparticles. *Materials Chemistry Horizons*, 2023. **2**(1): p. 49–56.
69. Hashim, M., et al., Investigation of structural, dielectric, magnetic and antibacterial activity of Cu-Cd-Ni-FeO_4 nanoparticles. *Journal of Magnetism and Magnetic Materials*, 2013. **341**: p. 148–157.
70. Mmelesi, O.K., et al., Cobalt ferrite nanoparticles and nanocomposites: Photocatalytic, antimicrobial activity and toxicity in water treatment. *Materials Science in Semiconductor Processing*, 2021. **123**: p. 105523.
71. kumar, N.H., D. Ravinder, and A. Edukondalu, Synthesis, structural, antimicrobial activity and dielectric properties of Ce^{3+}-doped Ni–Zn nano-ferrites. *Applied Physics A*, 2022. **128**(11): p. 978.
72. Alsafari, I.A., et al., Synthesis, characterization, photocatalytic and antibacterial properties of copper Ferrite/MXene ($CuFe_2O_4/Ti_3C_2$) nanohybrids. *Ceramics International*, 2021. **47**(20): p. 28874–28883.

7 Spinel and Inverse Ferrites for Energy Applications

Akhil Nigam, Sandeep Kaur and Navjeet Kaur

7.1 INTRODUCTION

In recent years, with the rapid development of industrial sectors, many challenges have been experienced by the global environment. This tends to lead to unprecedented destruction and discontinuity of power supply. Due to catalytic advanced techniques, environmental issues have been ameliorated, and environment-friendly energy has been developed for the consumer. Nowadays catalytic advanced techniques have penetrated numerous aspects of life, such as the production of clean energy, environmental rehabilitation, and sustainable development. In the past, spinel and inverse ferrites sometimes played a vital role in the production of electricity. These two types of magnetic materials have been used in various applications for energy production. Such materials exhibit some unique electric and magnetic properties which are available and valuable in a range of technical contexts [1]. Due to their excellent magnetic and catalytic properties, spinel and inverse ferrites have attracted the attention of authors and researchers. Comparison of Spinel and inverse spinel ferrites is given in Table 7.1.

However, in the last few years, the crisis of energy production and subsequent energy shortage has caused spinel and inverse ferrites to be introduced. These types of materials can be used in numerous applications for smooth function. Many authors or researchers have identified their applications in energy sectors comprising sensors, renewable energy sources, and energy storage devices.

Both spinel and inverse ferrites have been ascertained for their continuous contributions to energy production. Different strategies have been employed for the production of sustainable and flexible energy in various areas.

1. **Photovoltaic:** Solar energy plays an important role in energy production due to its low cost and low consumption. Spinel ferrite material can be preferred as photoanodes or photocathodes in photoelectrochemical cells. These are used for absorbing sunlight and separating water into oxygen and hydrogen. Hence, such material can be used as a clean fuel energy source.

DOI: 10.1201/9781032665658-7

TABLE 7.1
Difference between Spinel Ferrites and Inverse Ferrites

Aspects	Spinel Ferrites	Inverse Ferrites
Crystalline structure	It is in the form of a cubic.	It is available partially or inverted.
Magnetic characteristics	It is present in ferrimagnetic form.	It may be in nonmagnetic, ferromagnetic, or anti-ferromagnetic form.
Electric resistivity	It has low electric resistivity.	It has high electric resistivity.
Electrical and magnetic controlling	It covers limits for both types of controlling.	It covers more for both types of controlling.
Applications	It can be used in transformers, storage devices, and magnetic sensors.	It may be used in gas detection, capacitors, and sensors.

2. **Thermoelectric generator:** Using spinel ferrites with suitable doping exhibits an efficient high thermoelectric capability. These are used to convert waste energy into electricity. This helps in proper energy harvesting from industrial farms.
3. **Magnetization:** Inverse ferrite material can be used for the development of a magnetic energy generator, which converts mechanical energy into electrical energy by using the principle of electromagnetic induction.
4. **Catalysis:** Spinel ferrites have been used in terms of catalysis for numerous energy production applications. These may include the reduction of carbon dioxide (CO_2) gas and the reformation of methane (CH_4) gas.

The previous application of spinel and inverse ferrites was as magnets, but at that time, there was no further usage for a longer period. In the 1990s due to rapid development in advanced nanotechnology, spinel ferrites were revisited. Various authors have proposed these materials for energy applications. Their attention has increased toward catalytic applications of spinel ferrites and their derivatives in energy applications. Compared to large-sized spinel ferrite materials, nanomaterials are superior in thermal applications, mechanical applications, electrical applications, optical applications, and catalytic applications [2]. Therefore, most spinel ferrite materials can be further utilized and recycled by employing a magnet that supports the principle of green chemistry.

7.2 ROLE OF SPINEL FERRITES IN ENERGY APPLICATIONS

Ferrites have been observed and must be marked as remarkable materials made up of iron oxides with a combination of metal elements like zinc, cobalt, nickel, manganese, etc. All such types of metals are electroconductive in nature. Ferrites are of two types: soft nanoferrites and hard nanoferrites. The significance of using these materials is due to certain properties like high resistivity, low losses, and stability in temperature. The point of coercivity is higher in hard ferrites as compared to soft ferrites, so these materials can be used in electrical applications such as washing machines, communication systems, microwave absorbing systems, and switch-mode power supplies [3]. The presence of dopants in spinel ferrites may lead to site substitution. Hence, further doping may lead to the structural reversal of ferrites, and if the dopant differs from the metal in size configuration, it may cause deflection in the crystal structure.

Spinel ferrites are also considered magnetic spinel ferrites, and these have wide applications in the fabrication of electronic devices due to their electric and magnetic properties (Figure 7.1).

1. **Transformers:** Spinel ferrites are used for the construction of a transformer for the formation of a core. Using these materials they show high magnetic properties, establishing thermal stability and low losses.
2. **Microwave devices:** Spinel ferrite materials are used in applications such as isolators and conductors because of their low losses and high permeability. Further, they can be used in telecommunication systems and radar systems.
3. **Antenna devices:** For antenna designing, spinel ferrites are used to improve its performance by enhancing radiation characteristics.
4. **Power electronic devices:** Spinel ferrite materials can be used in power electronic devices such as inductors and chokes for managing and controlling power in electronic systems.

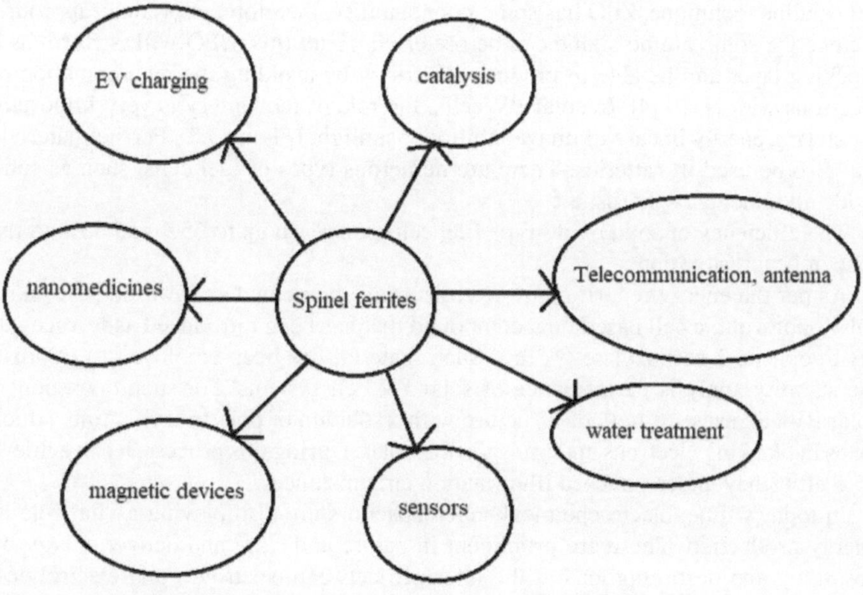

FIGURE 7.1 Applications of spinel ferrites.

5. **Charging of electric vehicles:** For charging of electric vehicles, spinel ferrite material can be used. The purpose of using these materials is to develop wireless charging through the optimization of energy transfer between electric vehicles and charging stations.
6. **Data storage devices:** For keeping data storage, hard disk drives, recording media, and magnetic tapes are utilized because they use spinel ferrites with thin film layers for storing data. These materials have more reliability in maintaining magnetic properties.
7. **Sensors:** In magnetic sensors, spinel ferrite materials are used. By the presence of the Hall effect principle, these can be used for detecting variations in magnetic fields and can convert them into electric fields.

7.3 ROLE OF SPINEL FERRITES IN RENEWABLE ENERGY SYSTEMS

With the presence of renewable energy systems (RESs), the problem of CO_2 gas emissions has been reduced and the efficiency of power plants has improved. Such environmental energy harvesting devices have garnered more interest due to the increasing reliable energy demand. There is a challenge for researchers to make eco-friendly energy production, so spinel ferrite materials have also been added to make it more suitable for energy applications. Different forms of renewable energy sources like solar (Photovoltaic) PV, wind turbines, geothermal, etc., have been introduced for the production of electricity in both rural and urban areas.

For renewable energy sources, solar cells are the most widely used effective source owing to their efficiency and ability to function in different places. Ferrite materials are enormously used for the production of solar cells. Perovskite bismuth ferrite (BFO) on zinc oxide-based solar PV cells has been introduced through the chemical solution technique. ZnO has some poor stability; therefore, to produce appropriate energy, some amino silane can be preferred. After this, BFO will perform as a blocking layer and be able to produce electricity by avoiding the recombination of electrons with N719 [4]. In solar PV cells, the role of the battery is very important for storing energy in case of unavailability of sunlight (Figure 7.2). Ferrite materials can also be used in batteries. There are numerous types of fuel cells, such as solid oxide alkaline, phosphorus, etc.

The efficiency of solid oxide-type fuel cells can reach up to 65% and 85% in the case of heat restoration.

As per the energy report, many scientists have fabricated an inverted perovskite solar photovoltaic cell based on a compound that has been introduced as ferrocenyl-*bis*-thiophene-2-carboxylate ($FcTc_2$). Such material has been employed to improve the stability analysis performance of solar PV cell systems. For such operation, a solar PV cell uses a p-n diode structure with a solution of poly(triaryl)amine, which helps in blocking electrons and transporting holes. During this process, it can achieve 25% efficiency under standard illumination circumstances.

In today's time, electrochemical supercapacitors are also playing a vital role in energy production. These are prominent in nature and store and deliver energy by absorbing and desorbing ions at the interface between electrodes and electrolytes.

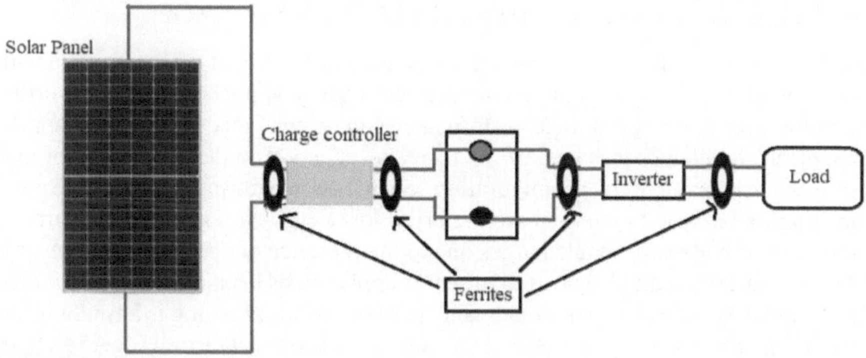

FIGURE 7.2 Role of ferrites in renewable energy sources.

Spinel and inverse-based ferrite materials are considered to be more predictable substitutes for electrode materials due to specific properties of supercapacitors like thermal conductivity, flexibility, reliability, and electrical conductivity. There are some wide extents of selecting supercapacitors such as $ZnMn_2O_4$, $CuCo_2O_4$, $MgCo_2O_4$, etc (Figure 7.3). So spinel and inverse ferrite materials play a vital role in supercapacitors, which are also known as ultra-capacitors. These supercapacitors are used for energy storage devices and can transmit energy whenever energy demand is required.

Many authors have investigated electrical conduction with the study of the bound of free ions in dielectrics and electronics. The author has performed an investigation by using ferrite materials such as $MgFe_2O_4$, which shows electrical conduction with increasing temperature, resulting in semiconductor properties.

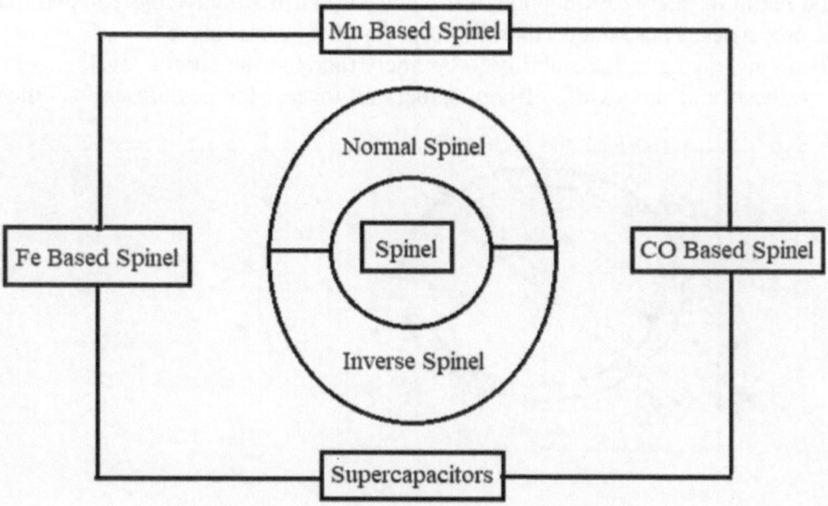

FIGURE 7.3 Role of ferrites in electrochemical supercapacitors [5].

7.4 ROLE OF SPINEL FERRITES FOR ENERGY STORAGE

Based on spinel ferrites under nanostructures, the availability of energy storage with a focus on electrocatalytic and electrochemical energy applications is noteworthy, Therefore, overall energy storage performance can be enhanced by choosing appropriate electrode materials. The performance also depends on the interfacing of ion-electrode considering porous configuration, ion diffusion pathways, and surface area. Transition metal-based materials are preferred for energy storage device electrodes as they can provide additional charges due to the presence of additional charges [6].

Ferrite materials can also be used in sensor applications because they have unique electric and magnetic properties that make them suitable for sensing information and transmitting it to another place. Spinel ferrites have a chemical formula of AFe_2O_4, in which A can be any divalent metal cation. This A can be selected as per the requirement of sensing data. Spinel ferrite materials are highly magnetic and susceptible to external magnetic excitation. These materials can exhibit magnetostrictive conditions where the dimensions of the material can be changed by an external magnetic field. Such types of materials can be used in magnetometers and magnetic sensors. Hall-type sensors also employ ferrite materials for detecting variations in magnetic field, which can be utilized in automobile applications such as anti-lock braking systems (Figure 7.4). In the case of inverse ferrite materials, they have the same chemical structure and can exhibit electric sensitivity [8–15]. Such types of characteristics make sensors more difficult to detect and can give stable responses. These can be employed for environmental observation and measurement of electric fields.

7.5 CONCLUSION

In conclusion, spinel ferrites and inverse ferrites have emerged as pivotal materials in the realm of energy production, attributed to their distinctive electric and magnetic properties. These materials exhibit exceptional performance in various energy applications, owing to their ability to be finely tuned at the atomic level. However, the synthesis and processing of both spinel and inverse ferrites present significant

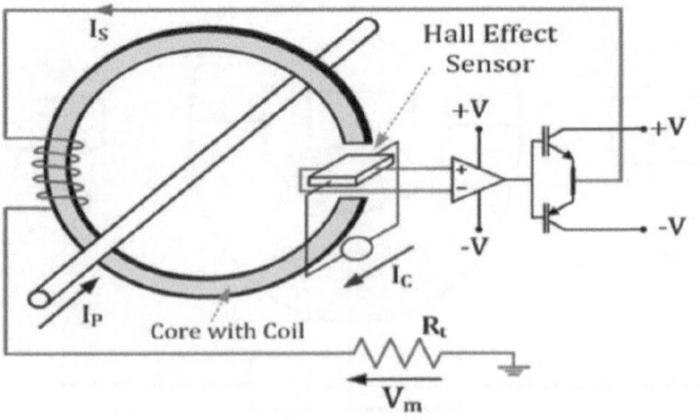

FIGURE 7.4 Role of ferrites in Hall effect sensor [7].

challenges. The procedures involved are often complex, requiring precise control over numerous parameters to optimize their characteristics fully. This complexity is further compounded by considerations of cost and material availability, which pose obstacles to their large-scale implementation in energy systems. Despite these challenges, the potential of ferrite materials in the energy sector is immense. Advances in nanotechnology offer promising avenues for the development of novel ferrite-based materials with enhanced properties. By leveraging innovative synthesis techniques and nanoscale engineering, it is possible to overcome some of the limitations associated with traditional ferrite materials. This could lead to the creation of next-generation ferrites with superior performance and efficiency, tailored specifically for advanced energy applications.

REFERENCES

1. H. Qin, Y. He, P. Xu, D. Huang, Z. Wang, H. Wang, Z. Wang, Y. Zhao, Q. Tian, and C. Wang, "Spinel ferrites(MFe_2O_4): Synthesis, improvement and catalytic application in environment and energy field," *Advances in Colloid and Interface Science*, vol. 294, pp. 1–22, 2021.
2. J. Deng, S. Feng, X. Ma, C. Tan, H. Wang, S. Zhou, T. Zhang, and J. Li, "Heterogeneous degradation of Orange II with preoxymonosulfate activated by ordered mesoporous $MnFe_2O_4$," *Separation and Purification Technology*, vol. 167, pp. 181–189, July 2016.
3. P. A. Vinosha, A. Manikandan, A. S. J. Ceicilia, A. Dinesh, G. F. Nirmala, A. C. Preetha, Y. Slimani, M. A. Almessiere, A. Baykal, and B. Xavier, "Review on recent advances of zinc substituted cobalt ferrite nanoparticles: synthesis characterization and diverse applications," *Ceramics Applications*, vol. 47, pp. 10512–10535, 2021.
4. S. Paul, B. Bharti, and Rajesh Kumar, "Role of ferrite materials in renewable energy harvesting," *Engineering Ferrites and Their Applications*, pp. 113–131, June 2023.
5. V. Jose, V. Jose, C. E. E. F. Christy, and A. S. Nesaraj, "Spinel-based electrode materials for application in electrochemical supercapacitors – present status and future prospectus," *Inorganic and Nano-Metal Chemistry*, vol. 52, no. 12, pp. 1449–1462, July 2021.
6. S. Jo, S. Pak, Y. W. Lee, S. Cha, J. Hong, and J. I. Sohn, "Enhancing the electrochemical energy storage performance of bismuth ferrite supercapacitor electrodes via simply induced anion vacancies," *International Journal of Energy Research*, vol. 1, pp. 1–9, 2023.
7. https://www.testandmeasurementtips.com/basics-of-hall-effect-sensing/
8. M. Kacki, M. S. Rylko, J. G. Hayes, and C. R. Sullivan, "A practical method to define high frequency electrical properties of MnZn ferrites," Proceedings IEEE Applied Power Electronics Conference and Exposition (APEC), New Orleans, LO, March 2020, pp. 216–222.
9. A. Stadler, M. Albach, and A. Bucher, "Calculation of core losses in toroids with rectangular cross section," Proceedings 12th International Power Electronics Motion Control Conference, Portoroz, Slovenia, August 2006, pp. 828–833.
10. S. Lin, T. Brinker, L. Fauth, and J. Friebe, "Review of dimensional resonance effect for high frequency magnetic components," Proceedings 21st Europen Conference Power Electronics Applied (EPE ECCE Eur.), Genova, Italy, September 2019, p. 1.

11. J. Zhang, Z. Wang, Q. Zhang, J. Wu, J. Tao, J. Chen, L. Jiang, L. Cao, "Enhancement of gyration effects by dysprosium-doped ferrite/piezoelectric magnetoelectric gyrators," *IEEE Transactions on Magnetics*, vol. 58, no. 2, pp. 1–5, February 2022.
12. S. Pavithra, R. Yuvakumar, A. Sakunthala, G. Ravi, and M. V. Reddy, "Spinnel ferrites for energy applications," *Ferrite Nanostructured Magnetic Materials Technologies and Applications*, pp. 723–732, 2023.
13. M. Kamran and M. Anis-ur-Rehman, "Influence of La^{3+} substitutions on structural, dielectric and electric properties of spinel cobalt ferrite," *Ceramic International*, vol. 49, no. 4, pp. 7017–7029, February 2023.
14. Md. Forhad Hussain, Md. Abdul Awal, Md. Johurul Islam, M.T. Khatun, S. Islam, and M. J. Rahman, "Synthesis and characterization of structural, magnetic and electrical properties of Mn substituted Co-Zn ferrite," *Journal of Magnetism and Magnetic Materials*, vol. 588, p. 171440, December 2023.
15. A. K. Pradhan, G. I. Choudhari and A. Datta, "Characteristics of cobalt ferrite nanoparticles for supercapacitor applications," 2023 8th International Conference on Computers and Devices for Communication (CODEC), Kolkata, March 2024.

8 Spinel and Inverse Ferrites for Electromagnetic Shielding

Madan Lal, Priyanka Thakur, Navdeep Sharma, and Fateh Singh Gill

8.1 INTRODUCTION

In the 21st century, the vast array of uses that ferrite-based materials offer has recently attracted the attention of researchers and academics. Usually referred to as iron oxides, they exhibit ferromagnetic behavior and contain one or more metals in their structure. They are divided into various groups according to their crystal structure and magnetic characteristics. Orthoferrite, hexaferrite, garnet, and spinel ferrite are among the well-known categories of ferrite. The typical formula for spinel ferrites is XY_2O_4, where X and Y are metal cations at octahedral and tetrahedral sites, respectively [1]. The chemical formula must contain at least one ferric atom (Fe^{3+}) for materials to qualify as spinel ferrites. The formula MFe_2O_4, where M can be any metal with an oxidation state of +2 (e.g., Cu^{2+}, Fe^{2+}, Ni^{2+}, Mg^{2+}, Mn^{2+}, Zn^{2+}, Co^{2+}, etc.), is the best representation of spinel ferrites [2–5]. Because of their magnetic properties, exceptional chemical and thermal stability, they exhibits several uses including medication delivery, magnetic resonance imaging (MRI), hyperthermia cancer treatment, anode materials for Li-ion batteries, microwave devices, ferrofluids, supercapacitors, magnetic refrigeration (MR), gas sensors, solar cells, microwave absorbers, in electromagnetic shielding, and in the paint industry [6–9].

The cubic unit cell of spinel ferrites consists of 32 octahedral and 64 tetrahedral interstices between closely spaced oxygen atoms. The unit cell is composed of eight formula units ($M_8Fe_{16}O_{32}$), in which eight cations occupy 64 tetrahedral interstices (A sites) and 16 cations occupy 32 octahedral interstices (B sites). Thirty-two oxygens create a densely packed face-centered cubic (fcc) structure. In the spinel unit cell, a total of 24 cations (64 tetrahedral and 24 octahedral) occupy the 96 available interstices. A small number of unoccupied interstitial sites also allow for cation transit between the sites. Spinel ferrites are classified as normal, inverse, or mixed varieties depending on where the +2 and +3 cations are located in the crystal lattice.

DOI: 10.1201/9781032665658-8

In typical spinel ferrites, the M^{2+}-ions are found at the tetrahedral sites, whereas the Fe^{3+}-ions are found at the octahedral sites. $ZnFe_2O_4$, $MgFe_2O_4$, $CoFe_2O_4$, and $CdFe_2O_4$ are examples of conventional spinel ferrites. In the inverse spinel ferrites, M^{2+} ions only occupy the octahedral sites, whereas Fe^{3+} ions occupy both the tetrahedral and octahedral sites. In mixed spinel ferrites, M^{2+} and Fe^{3+} ions alternately occupy the tetrahedral and octahedral sites [10–13].

Human life has been made easier by the widespread development and implementation of electronic applications, wireless communication systems, and tools in civil and military fields. However, this heavy reliance on electronic gadgets and instruments has grown and expanded air pollutant known as unwanted electromagnetic radiation. An electromagnetic wave is the synchronized movement of two perpendicular magnetic and electric fields. Due to electromagnetic interferences (EMI), the undesired electromagnetic waves produced by electronic, electrical, or hybrid equipment and instruments interfere with the operation and performance of nearby devices and instruments [14–17]. In addition to efficiency, the durability and safety of commercial and military electronic equipment can be negatively affected. Apart from the malfunctioning of electronic devices and appliances, this EMI issue also has an impact on people's health [18–20]. Exposure to undesirable electromagnetic radiation over an extended period raises the risk of developing cancer, migraines, heart issues, asthma, childhood leukemia, and even miscarriage. Therefore, good electromagnetic wave shielding or absorbent materials need to be developed to overcome the effect of EMI [21–24]. Due to intriguing uses in the low- and high-frequency ranges, spinel ferrite nanoparticles have gained a lot of attention in a variety of scientific fields in recent years. Spinel ferrite nanoparticles are used as microwave absorbers or electromagnetic wave shields in one of the most recent high-frequency applications. Spinel ferrite nanostructures are also attracting more attention than sophisticated nanocomposites due to their exceptional electromagnetic interference shielding or microwave absorption characteristics [25–28]. For electromagnetic shielding, a variety of high-performance spinel ferrite nanocomposites have been created, including spinel ferrites based on polymers, metal, metal-oxide, carbon, rubber, and cement or concrete [29–32]. Understanding the electromagnetic wave shielding mechanism and related electromagnetic properties and parameters is crucial to comprehending how these advanced spinel ferrite nanostructures and nanocomposites work. Therefore, this chapter fully covers the mechanism, properties of spinel ferrites, and other parameters related to EMI.

8.2 MECHANISM OF ELECTROMAGNETIC SHIELDING

In addition to grounding, filtering, and bonding procedures, EMI shielding is another way to achieve electromagnetic compatibility (EMC). This idea suggests building enclosures with the right kind of material, structure, and shape to change the way that most of the unwanted electromagnetic radiation enters or exits the equipment. By passing the electromagnetic wave through conductive or ferromagnetic materials, the path is changed through absorption or reflection. Magnetic and electric components both make up an electromagnetic wave. They are perpendicular to one another and have the same frequency of travel. Electric components are blocked by conductive materials, while magnetic components are blocked by materials with high magnetic permeability. It is sufficient to safeguard one component because an electromagnetic wave cannot exist with just one component. Different strategies are used to filter each out when it comes to EMI shielding [33–36]. In general, the ratio of incident to transmitted energy, or the electromagnetic field, is known as the material's total shielding efficiency (S_T), and it is stated as [37]:

$$S_T = 10\log\left(\frac{P_{\text{input}}}{P_{\text{output}}}\right) = 20\log\left(\frac{E_{\text{input}}}{E_{\text{output}}}\right) = 20\log\left(\frac{H_{\text{input}}}{H_{\text{output}}}\right)\text{dB} \tag{8.1}$$

The subscripts "input" and "output" refer to incident and transmitted waves, respectively, while dB stands for decibels (the measure of a material's capacity to attenuate electromagnetic waves as they travel through it). Here, P stands for power, E for the electric field, and H for the magnetic field.

The S_T for a material is equal to the total of the electromagnetic wave attenuation from the shielding mechanisms of absorption (S_A), reflection (S_R), and multiple reflections (S_M), according to Schelkunoff's theorem [38–41] (as shown in Figure 8.1):

$$S_T = S_A + S_R + S_M \tag{8.2}$$

Electromagnetic shielding performance values of some spinel, inverse, and its nanocomposites summarized in Table 8.1 according to reports from the literature.

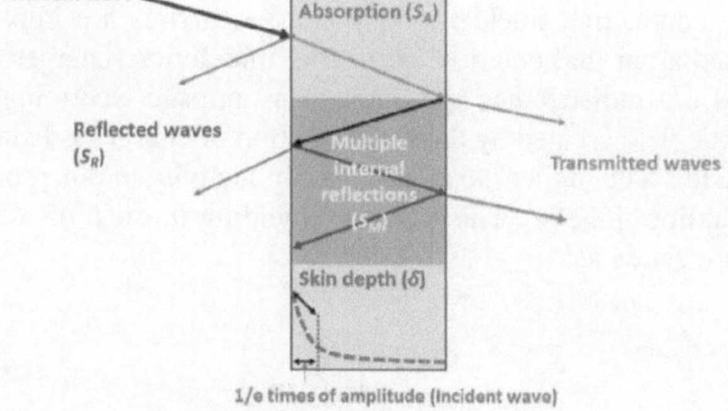

FIGURE 8.1 Potential interactions between materials and electromagnetic waves.

TABLE 8.1
Electromagnetic Shielding Performance of Some spinel Ferrites, Inverse Spinel Ferrites, and Its Nanocomposites

Materials	Maximum R_L (-dB)	Frequency at Maximum R_L (GHz)	Skin Depth δ (mm)	Ref.
Fe_3O_4	45.2	4.67	–	[42]
$NiFe_2O_4$	2.4	6.7	5.5	[43]
$CoFe_2O_4$	55	9.25	2	[44]
3wt% MWCNT+3wt% Fe_3O_4	31	18	2.1	[6]
3wt% MWCNT+3wt% $CoFe_2O_4$	26	18	2.9	
3wt% MWCNT+3wt% $NiFe_2O_4$	24	18	3.4	
Cubic $NiFe_2O_4$/GN-PANI	50.5	12.5	2.5	[45]
$CoFe_2O_4$/GO	10	9.8	0.5	[46]
GO+$NiFe_2O_4$	46	7	0.5	[47]
rGO-Fe_3O_4	38	18	0.91	[48]
$Co_{0.33}Zn_{0.67}Fe_2O_4$-TPU	67.8	12.4	0.80	[49]
Polypyrrole/BST/rGO/Fe_3O_4	48.6	10	2.5	[50]
PTH/$Ni_{0.5}Zn_{0.5}Fe_{2x}Ce_xO_4$	34	10	2.0	[51]
rGO-$MnFe_2O_4$ + MWCNTs in PVDF	38	2–18	5.5	[48]
MWNTs + rGO-Fe_3O_4 in polycarbonate (PC)/ poly(styrene-*co*-acrylonitrile) (SAN) blend	50.7	8–18	0.92	[52]

8.2.1 Shielding through Reflection

The main technique for EMI shielding is reflection, which reduces the electric component of the EMI. For the material to create EMI reflection, mobile charge carriers are required. The conductive shield's mobile charge carriers are impacted by the electromagnetic radiation that enters. Due to this interaction, charges move along the conductor and are redistributed, producing an opposite electromagnetic field. The electromagnetic field created by the redistribution of charges is balanced by the magnetic field outside. The higher the material's conductivity in this process the better its shielding qualities [53, 54]. The effective shielding through the reflection of a material can be calculated as:

$$S_R = -10\log_{10}\left(\frac{\sigma_T}{16 f \epsilon \mu r}\right) \tag{8.3}$$

where σ_T = total conductivity (S/cm), f = frequency of the electromagnetic waves, μ_r = relative magnetic permeability, and ε = the electrical permittivity of the medium, respectively [55].

The drawback of this mechanism is that it loses its shielding abilities if there is a discontinuity in the enclosure that is larger than the wavelength of the external electromagnetic field. As a result, the enclosure design's hole and opening sizes are kept to a minimum. For higher electromagnetic wave frequencies, this is not feasible. Filtering equipment is the sole technique to combat this in high-frequency EMI.

The skin effect, which can be observed in AC circuits, is yet another issue. Charges build up at the surface or topmost layers of the conductor as AC passes through it, increasing the current density there. Less utilization of the interior part affects conductivity and, eventually, shield performance. When it comes to high-frequency electromagnetic radiation, this effect is noticeable. The conductor's surface area can be increased, improving the effective conducting cross-section as a result. Another option is to electroplate the surface with a substance that conducts electricity very well, such as silver [6].

8.2.2 Shielding through Absorption

This is the EMI shielding secondary mechanism that affects the magnetic EMI component. The substance must possess both electric and magnetic dipoles to achieve EMI absorption. These are materials with a high magnetic permeability and dielectric constant. The magnetic field lines are cut in the presence of an external magnetic field because they prefer to pass through the material. By creating a channel within itself, an enclosure with this ability absorbs the magnetic and electric field lines. However, the low conductivity of these materials presents a challenge when using them. They are therefore less effective at shielding against the electromagnetic wave's electric component. Eddy currents play a role in the absorption process by diminishing the incoming electromagnetic wave. When the electromagnetic wave oscillates at a high frequency and causes currents to flow inside the conductor, this is what is seen. The magnetic field produced by the eddy currents opposes the magnetic field outside. Eddy currents are amplified when materials with high electric conductivity are used. The effective shielding can be calculated as in Equation (8.4):

$$S_A = -8.68t\left(\frac{\sqrt{f\mu_r\sigma_T}}{2}\right) \tag{8.4}$$

where t = thickness of the shielding material. Equation (8.4) shows that the S_A rises as the frequency increases. Further, the S_A is directly proportional to the shielding material's and μ_r. Thus, the S_A depends on the thickness (t), magnetic permittivity (μ_r), and electrical conductivity (σ_T) of the shielding materials. The shielding materials must show both electric and magnetic dipoles to absorb electromagnetic waves. The shield current (ohmic losses) and magnetic hysteresis losses are the main causes of the absorption [56].

8.2.3 Shielding through Multiple Reflections

Higher frequency electromagnetic waves can only partially pass through a conducting shield at its surface, and the strength of the field decreases exponentially with shield thickness. The skin depth (δ) is defined (shown in Figure 8.1) as the distance (or thickness of the shield) at which the electromagnetic wave is reduced by 1/*e*, or 37%, and is expressed mathematically as:

$$\delta = \frac{1}{\sqrt{\pi f \mu \sigma}} \tag{8.5}$$

The reflected wave becomes reflected and bounces back and forth between the first and second borders of a thin shield material. The following is an explanation of how effective shielding against multiple reflections works:

$$S_M = 20 \log 10\left(1 - e^{\frac{2t}{\delta}}\right) = 20 \log 10\left(1 - e^{\frac{E_A}{10}}\right) \tag{8.6}$$

It is obvious from Equation (8.5) that S_M is quite low for materials with considerable absorption. Therefore, the value of S_M is minimal and typically left out of the equation for the overall sheathing effectiveness (S_T) for a thick shielding material with high S_A[57].

8.3 PROPERTIES RELATED TO ELECTROMAGNETIC SHIELDING MECHANISMS

The complex permittivity and complex permeability of electromagnetic interference shielding materials are related to their basic shielding processes

8.3.1 Reflection Loss

The microwave absorption of electromagnetic interference shielding materials is caused by the reflection loss (R_L). According to the transmission line theory, the reflection loss (R_L) can be calculated using the following equation:

$$R_L\left(\text{dB}\right) = 20\log 10\left(\frac{|Z_{in} - Z_0|}{|Z_{in} + Z_0|}\right) \tag{8.7}$$

$$S_A = -R_L\left(\text{dB}\right) \tag{8.8}$$

where Z_{in} = normalized electromagnetic wave absorber material input impedance, = free space impedance, (here $Z_0 = \left(\frac{\mu_0}{\varepsilon_0}\right)^2$; the permittivity and permeability in open space are denoted by $\varepsilon_0$and μ_0. When the absorber material is supported by a metallic reflector, its normalized input impedance is given by

$$Z_{in} = \left(\sqrt{\frac{\mu_r}{\varepsilon_r}}\right)\tanh\left(j\frac{2\pi}{c}fd\right) \tag{8.9}$$

where d is the thickness of the electromagnetic absorber material, c is the light velocity, and ε_r is the complex permittivity and permeability of the electromagnetic wave absorber material [58–61].

8.3.2 Dielectric Loss

The complex permittivity of any material is written as:

$$\varepsilon_r = \varepsilon' - i\varepsilon'' \tag{8.10}$$

where i = imaginary number, ε_r= relative permittivity, ε' = real part of permittivity (it represents the dielectric constant, charge storage, or electric energy storage capacity), and ε'' = the dielectric losses or dissipation represented by the imaginary fraction of permittivity. Thus, the dielectric loss tangent ($tan\delta_\varepsilon$) can be used to determine the magnitude of dielectric losses [62, 63].

$$tan\delta_\varepsilon = \frac{\varepsilon''}{\varepsilon'} \tag{8.11}$$

8.3.3 Magnetic Loss

A material's magnetic permeability can be stated as follows:

$$\mu_r = \mu' - i\mu'' \tag{8.12}$$

where μ_r= relative permeability, μ' = imaginary component of permeability, which indicates the magnetic losses or dissipation, and i = real part of permeability, which denotes the magnetic storage or magnetic energy storage capacity/magnetic moment. Thus, the magnetic loss tangent ($tan\delta_\mu$) can be used to determine the magnitude of dielectric losses as follows:

$$\tan\delta_\mu = \frac{\mu''\ \mu''}{\mu'\ \mu'} \tag{8.13}$$

Consequently, the geometry, form, conductance, and type of the incident radiation exhibit complicated dependencies on the electromagnetic interference shielding material's efficacy in shielding [64].

8.3.4 Dielectric Loss Source

As per the Debye theory, the potential of electromagnetic energy is lost, and this is shown by the fictitious permittivity (ε''). The dielectric loss tangent ($tan\delta$) provides insight into the contributions of both polarization loss and conduction loss.

$$\varepsilon' = \varepsilon_{\infty} + \frac{\varepsilon_s - \varepsilon_{\infty}}{1+\omega^2\tau^2} \tag{8.14}$$

$$\varepsilon'' = \frac{\varepsilon_s - \varepsilon_{\infty}}{1+\omega^2\tau^2}\omega\tau + \frac{\sigma}{\omega\varepsilon_0} \tag{8.15}$$

where ω is the angular frequency, σ is the electrical conductivity, τ is the relaxation polarization time, and ε_s is the static dielectric constant. ε_{∞} is the relative dielectric constant at the high-frequency limit. The loss of polarization results from two processes: dipole orientation, which is generated by bound charges, and electronic, ionic, and interfacial polarization, which is caused by the trapping of space charge. Electronic and ionic polarization can be ignored in the low microwave frequency range because they only contribute at high frequencies (above 1000 GHz). Dipole polarization develops because of the material's imperfections and residual groups, which are mostly dependent on the preparation process and other preparation-related factors, etc. The cause of trapped charges at the interfaces is interfacial polarization and related relaxation [65, 66].

According to Cole-Cole, equation given in Equation (8.15) [25], which is based on Equations (8.13) and (8.14):

$$\left(\varepsilon' - \frac{\varepsilon_s - \varepsilon_{\infty}}{1+\omega^2\tau^2}\right)^2 + \left(\varepsilon''\right)^2 = \left(\frac{\varepsilon_s - \varepsilon_{\infty}}{2}\right)^2 \tag{8.16}$$

where ε_s and ε_{∞} corresponding to the higher frequency static and relative dielectric permittivity.

The ε'' vs. ε' plot will display a semicircle equal to one Debye relaxation process if polarization relaxation occurs. The Cole-Cole semicircle plot is the name of the plot. Semicircles that are superimposed on top of one another denote that each semicircle is connected to the polarization and mobility of conducting electrons in the heterogeneous interface, which affects permittivity [67, 68]. Based on the free-electron theory, it is possible to explain the dielectric loss, which is mostly caused by conduction loss [69]:

$$\varepsilon'' = \frac{\sigma}{2\pi f \varepsilon_0} \tag{8.17}$$

where $\varepsilon_0 = 8.854 \times 10^{-12}$ $C^2/N.m^2$ (electrical permittivity in vacuum), f = applied frequency, and σ = electrical conductivity.

8.3.5 Magnetic Loss Source

Eddy current loss, exchange resonance, natural resonance, and domain wall resonance are frequently the causes of magnetic loss. Domain walls will vibrate at their equilibrium position, parallel to the plane of the sample material, under an AC magnetic field $H = H_0 e^{i\omega t}$. A relaxation-type permeability spectrum, sometimes referred to as a Debye-type spectrum, is frequently seen in domain wall resonance. A significant amount of energy is absorbed from the static magnetic field when the energy of the AC transverse field equals the precessional frequency. Natural resonance is associated with the magnetization vector that exists in the static magnetic field. Griffith first discovered what is now known as ferromagnetic resonance in 1946 [70–73]. When ferromagnetic resonance occurs in the absence of an applied magnetic field, it is commonly referred to as natural resonance. The ferromagnetic resonance hypothesis states that the following connection can explain the natural resonance frequency (f_r):

$$f_r = \frac{\gamma}{2\pi} H_{eff} \tag{8.18}$$

where $\frac{\gamma}{2\pi}$ = gyromagnetic ratio, H_{eff} = effective magnetic field (here, $H_{eff} = \frac{4|K_1|}{3\mu_0 M_s}$; K_1= magnetocrystalline anisotropic coefficient for magnetic material; saturation magnetization is represented by $3\mu_0 M_s$. The increased anisotropic energy is particularly advantageous for the improved shielding and absorption of electromagnetic waves [74–76].

The magnetic loss caused by eddy current loss can be expressed using the following relationship:

$$C_0 = \frac{2\pi\sigma\mu_0 d^2}{3} \tag{8.19}$$

where d is the diameter of the nanoparticles, σ is electrical conductivity, and μ_0 is vacuum permeability. According to the skin effect criterion, if eddy current loss is the cause of magnetic loss, then the C_0 value stays constant as the frequency changes [77]. The exchange resonance between graphene capsules and Fe_3O_4 nanocrystals was discovered by Jian et al. in 2016 [71]. Remarkably, the composite system's negative imaginary permeability showed that the motion of charges within the three-dimensional hollow graphene capsules is what emits magnetic energy. The Maxwell equations state that an alternating electric field and a magnetic field are produced when charges move in an electromagnetic environment. Exchange resonances are generally found at higher frequencies than natural resonances, as stated by the Aharoni theory, and this has been shown in ferromagnetic nanoparticles. Based on the exchange resonance mode, the resonance frequency can be written as follows:

$$\frac{\omega}{\gamma_0} = \frac{C\mu_{kn}^2}{R^2 M_s} + H_0 - \frac{4\pi}{3} M_s + \frac{2k_1}{M_s} \tag{8.20}$$

where C = exchange coefficient, R = radius of spherical particles, H_0 = applied DC field, K_1 = magnetocrystalline anisotropy field, and μ_{kn} are the eigenvalues, which are the roots of the equation:

$$\left| \frac{dj_n\left(\frac{\mu_{kn}}{R}\right)}{d_r} \right|_{r=R} = 0 \tag{8.21}$$

where dj_n = spherical Bessel function. Equation (8.19) makes it clear that the exchange resonance frequency is connected with μ_{kn} squared, which explains why it appears in the higher frequency region [78–81].

8.4 ELECTROMAGNETIC SHIELDING MECHANISMS' GOVERNING PARAMETERS

The two key variables in figuring out a material's characteristics related to microwave absorption and electromagnetic interference shielding are the attenuation constant and impedance matching. The shielding material's attenuation characteristics are defined by the attenuation constant. On the other hand, impedance matching reveals the electromagnetic wave's propagation level into the shielding material [82 –84].

8.4.1 Matching Impedance Condition

By matching the impedance, the adoption of electromagnetic waves can be controlled. A favorable balance between the dielectric loss tangent (tan δ_ε) and the magnetic loss tangent (tan δ_μ) can be attributed to the improved electromagnetic wave adoption accomplished by good impedance matching. Based on Equations (8.10) and (8.12), the ideal electromagnetic matrix for zero reflection of the electromagnetic wave should satisfy the condition tan δ_ε = tan δ_μ. However, the materials that are now being researched are unable to maintain tan δ_ε = tan δ_μ across the entire frequency range, which implies that the absorber material's electromagnetic wave absorption is frequency dependent. The following relationship can be used to obtain the impedance matching coefficient:

$$Z = \frac{z_1}{Z_0} \tag{8.22}$$

$$Z_1 = \left(\frac{\mu_r}{\epsilon_r}\right)^{1/2} \left(\frac{\mu_r}{\epsilon_r}\right)^{1/2} Z_0 \tag{8.23}$$

$$|\varepsilon| = (\varepsilon' + \varepsilon'')^{1/2} \tag{8.24}$$

$$|\mu| = (\mu' + \mu'')^{1/2} \tag{8.25}$$

where Z_0 = impedance in free space, Z_1 = impedance of shielding material, and μ_r, ε_r are the relative permeability and permittivity of the shielding, respectively. The stronger impedance matching coefficient indicates that the shielding material is exhibiting improved impedance matching. Additionally, the amount of electromagnetic waves that propagates through shielding materials is determined by impedance matching [85–87].

8.4.2 Attenuation Constant

The study of the attenuation constant (α), which denotes the shielding material's capacity for internal attenuation, is crucial for gaining a deeper understanding of the electromagnetic interference absorption mechanism. The following formula can be used to determine the attenuation constant:

$$\alpha = \frac{\sqrt{2\pi f}}{c}\sqrt{(\mu''\varepsilon'' - \mu'\varepsilon')\sqrt{(\mu''\varepsilon'' - \mu'\varepsilon') + (\mu'\varepsilon'' - \mu''\varepsilon')^2}} \tag{8.26}$$

where, correspondingly, c = light velocity and f = electromagnetic wave propagation frequency. Equation (8.25) suggests that high values of the imaginary parts of permeability and permittivity may lead to a larger attenuation constant (α). A high attenuation constant value can be found using a high dielectric constant and magnetic loss. However, a high value for the attenuation constant does not necessarily translate into high magnetic loss or dielectric loss features [88–90].

8.5 CONCLUSION

This chapter presents the use of spinel, inverse ferrite nanoparticles, and their nanocomposites as electromagnetic interference shielding materials. It also emphasizes the basic ideas required to understand and develop high-performance shielding against electromagnetic interference and materials for microwave absorption The relationship between electromagnetic properties, such as permittivity, permeability, electrical conductivity, dielectric loss, magnetic loss, relaxations, and so on, and shielding and absorbing materials has been discussed. Furthermore, there is a focus on the electromagnetic parameters – such as impedance matching and the attenuation constant – that regulate the shielding process against electromagnetic interference.

REFERENCES

1. Thakur, P., Thakur, P., Kishore, K., Singh, M., Sharma, S., Sharma, P., . . . Lal, M. (2023). Structural, morphological, and magnetic properties of $CoFe_2O_4$ nano-ferrites synthesized via Co-precipitation route. *Materials Today: Proceedings.*
2. Wang, Y., Wang, W., Zhu, M., & Yu, D. (2017). Electromagnetic wave absorption polyimide fabric prepared by coating with core–shell $NiFe_2O_4$@ PANI nanoparticles. *RSC Advances, 7*(68), 42891–42899.
3. Gandhi, N., Singh, K., Ohlan, A., Singh, D., & Dhawan, S. (2011). Thermal, dielectric and microwave absorption properties of polyaniline–$CoFe_2O_4$ nanocomposites. *Composites Science and Technology, 71*(15), 1754–1760.
4. Kalidass, J., & Sivasankar, T. (2023). Mesoporous core/shell MnFe2O4 nanocomposite derived from facile sonoelectrochemical process: An eco-friendly method for rapid synthesis and versatile industrial applications. *Journal of the Taiwan Institute of Chemical Engineers, 144*, 104766.
5. Ramadan, R., Uskoković, V., & El-Masry, M. M. (2023). Triphasic $CoFe_2O_4$/$ZnFe_2O_4$/$CuFe_2O_4$ nanocomposite for water treatment applications. *Journal of Alloys and Compounds, 954*, 170040.
6. Biswas, S., Panja, S. S., & Bose, S. (2018). Physical insight into the mechanism of electromagnetic shielding in polymer nanocomposites containing multiwalled carbon nanotubes and inverse-spinel ferrites. *The Journal of Physical Chemistry C, 122*(34), 19425–19437.
7. Thakur, P., Gupta, K., Thakur, P., Kumar, A. S., Sudarsanan, V., Sharma, P., & Lal, M. (2023). Improvement in the structural, dielectric, and magnetic properties of CFO-doped KNNS-BKT ceramics. *Journal of Materials Science: Materials in Electronics, 34*(4), 311.
8. Amiri, M., Salavati-Niasari, M., & Akbari, A. (2019). Magnetic nanocarriers: Evolution of spinel ferrites for medical applications. *Advances in Colloid and Interface Science, 265*, 29–44.
9. Thieu, Q. Q. V., Kidanu, W. G., Nguyen, H. D., Nguyen, T. L. T., Le, M. L. P., Nguyen, D. Q., … Nguyen, T. L. (2022). Spinel Ni-ferrite advanced high-capacity anode for Li-ion batteries prepared via coprecipitation route. *Ceramics International, 48*(21), 31470–31477.
10. Reddy, D. H. K., & Yun, Y.-S. (2016). Spinel ferrite magnetic adsorbents: Alternative future materials for water purification? *Coordination Chemistry Reviews, 315*, 90–111.
11. Masunga, N., Mmelesi, O. K., Kefeni, K. K., & Mamba, B. B. (2019). Recent advances in copper ferrite nanoparticles and nanocomposites synthesis, magnetic properties and application in water treatment. *Journal of Environmental Chemical Engineering, 7*(3), 103179.
12. Šutka, A., & Gross, K. A. (2016). Spinel ferrite oxide semiconductor gas sensors. *Sensors and Actuators B: Chemical, 222*, 95–105.
13. Zhang, R., Qin, C., Bala, H., Wang, Y., & Cao, J. (2023). Recent Progress in Spinel Ferrite (MFe_2O_4) Chemiresistive Based Gas Sensors. *Nanomaterials, 13*(15), 2188.
14. Vyas, M. K., & Chandra, A. (2016). Ion–electron-conducting polymer composites: Promising electromagnetic interference shielding material. *ACS Applied Materials & Interfaces, 8*(28), 18450–18461.
15. Biswas, S., Kar, G. P., & Bose, S. (2015). Tailor-made distribution of nanoparticles in blend structure toward outstanding electromagnetic interference shielding. *ACS Applied Materials & Interfaces, 7*(45), 25448–25463.

16. Manna, K., & Srivastava, S. K. (2017). Fe_3O_4@ carbon@ polyaniline trilaminar core–shell composites as superior microwave absorber in shielding of electromagnetic pollution. *ACS Sustainable Chemistry & Engineering, 5*(11), 10710–10721.
17. Hua, T., Guo, H., Qin, J., Wu, Q., Li, L., & Qian, B. (2022). 3D printing lamellar $Ti_3C_2T_x$ MXene/graphene hybrid aerogels for enhanced electromagnetic interference shielding performance. *RSC Advances, 12*(38), 24980–24987.
18. Bayat, M., Yang, H., Ko, F., Michelson, D., & Mei, A. (2014). Electromagnetic interference shielding effectiveness of hybrid multifunctional Fe_3O_4/carbon nanofiber composite. *Polymer, 55*(3), 936–943.
19. Biswas, S., Panja, S. S., & Bose, S. (2017). Unique multilayered assembly consisting of "flower-like" ferrite nanoclusters conjugated with MWCNT as millimeter wave absorbers. *The Journal of Physical Chemistry C, 121*(26), 13998–14009.
20. Zhang, C., Ji, C., Park, Y. B., & Guo, L. J. (2021). Thin-metal-film-based transparent conductors: Material preparation, optical design, and device applications. *Advanced Optical Materials, 9*(3), 2001298.
21. Kheifets, L., Afifi, A. A., & Shimkhada, R. (2006). Public health impact of extremely low-frequency electromagnetic fields. *Environmental Health Perspectives, 114*(10), 1532–1537.
22. Woo, S., Kim, G., You, J., Park, S., Bae, J., & Lim, B. (2023). Formation of silver layer with a multiporous structure from silver nanoparticles for highly efficient electromagnetic interference shielding materials. *ACS Applied Nano Materials, 6*(12), 10087–10096.
23. Zhang, P., Cao, Z., Liu, C., Li, P., Kong, H., Li, T., … Xu, R. (2023). Ultra-thin freestanding graphene films for efficient thermal insulation and electromagnetic interference shielding. *RSC Advances, 13*(28), 19388–19402.
24. Sezer Hicyilmaz, A., & Celik Bedeloglu, A. (2022). Electromagnetic interference shielding thermoplastic composites reinforced with carbon based hybrid materials: A review. *Composite Interfaces, 29*(13), 1413–1470.
25. Li, N., Huang, G.-W., Li, Y.-Q., Xiao, H.-M., Feng, Q.-P., Hu, N., & Fu, S.-Y. (2017). Enhanced microwave absorption performance of coated carbon nanotubes by optimizing the Fe_3O_4 nanocoating structure. *ACS Applied Materials & Interfaces, 9*(3), 2973–2983.
26. Dalal, M., Greneche, J.-M., Satpati, B., Ghzaiel, T. B., Mazaleyrat, F., Ningthoujam, R. S., & Chakrabarti, P. K. (2017). Microwave absorption and the magnetic hyperthermia applications of $Li_{0.3}Zn_{0.3}Co_{0.1}Fe_{2.3}O_4$ nanoparticles in multiwalled carbon nanotube matrix. *ACS Applied Materials & Interfaces, 9*(46), 40831–40845.
27. Zeng, X., Jiang, G., Zhu, L., Wang, C., Chen, M., & Yu, R. (2019). Fe_3O_4 nanoflower-carbon nanotube composites for microwave shielding. *ACS Applied Nano Materials, 2*(9), 5475–5482.
28. Tian, X., Xu, X., Bo, G., Su, X., Yan, J., & Yan, Y. (2022). A 3D flower-like Fe_3O_4@PPy composite with core-shell heterostructure as a lightweight and efficient microwave absorbent. *Journal of Alloys and Compounds, 923*, 166416.
29. Zhou, W., Hu, X., Bai, X., Zhou, S., Sun, C., Yan, J., & Chen, P. (2011). Synthesis and electromagnetic, microwave absorbing properties of core–shell Fe_3O_4–poly(3,4-ethylenedioxythiophene) microspheres. *ACS Applied Materials & Interfaces, 3*(10), 3839–3845.
30. Liu, Y., Chen, Z., Zhang, Y., Feng, R., Chen, X., Xiong, C., & Dong, L. (2018). Broadband and lightweight microwave absorber constructed by in situ growth of hierarchical $CoFe_2O_4$/reduced graphene oxide porous nanocomposites. *ACS Applied Materials & Interfaces, 10*(16), 13860–13868.

31. Wang, Y., Huang, Y., & Wang, Q. (2012). Preparation and magnetic properties of $BaFe_{12}O_{19}/Ni_{0.8}Zn_{0.2}Fe_2O_4$ nanocomposite ferrite. *Journal of Magnetism and Magnetic Materials, 324*(19), 3024–3028.
32. He, Y., Lu, L., Sun, K., Wang, F., & Hu, S. (2018). Electromagnetic wave absorbing cement-based composite using Nano-Fe_3O_4 magnetic fluid as absorber. *Cement and Concrete Composites, 92*, 1–6.
33. Wang, H., Li, S., Liu, M., Li, J., & Zhou, X. (2021). Review on shielding mechanism and structural design of electromagnetic interference shielding composites. *Macromolecular Materials and Engineering, 306*(6), 2100032.
34. Wang, M., Tang, X.-H., Cai, J.-H., Wu, H., Shen, J.-B., & Guo, S.-Y. (2021). Construction, mechanism and prospective of conductive polymer composites with multiple interfaces for electromagnetic interference shielding: A review. *Carbon, 177*, 377–402.
35. Al-Saleh, M. H., & Sundararaj, U. (2009). Electromagnetic interference shielding mechanisms of CNT/polymer composites. *Carbon, 47*(7), 1738–1746.
36. Zhang, L., Bi, S., & Liu, M. (2018). Lightweight electromagnetic interference shielding materials and their mechanisms. *Electromagnetic Materials and Devices*, 1–10.
37. Gupta, S., & Tai, N.-H. (2019). Carbon materials and their composites for electromagnetic interference shielding effectiveness in X-band. *Carbon, 152*, 159–187.
38. Wanasinghe, D., & Aslani, F. (2019). A review on recent advancement of electromagnetic interference shielding novel metallic materials and processes. *Composites Part B: Engineering, 176*, 107207.
39. Kumar, P., Narayan Maiti, U., Sikdar, A., Kumar Das, T., Kumar, A., & Sudarsan, V. (2019). Recent advances in polymer and polymer composites for electromagnetic interference shielding: Review and future prospects. *Polymer Reviews, 59*(4), 687–738.
40. Schelkunoff, S. (1936). Some equivalence theorems of electromagnetics and their application to radiation problems. *The Bell System Technical Journal, 15*(1), 92–112.
41. Schelkunoff, S. (1956). *Electromagnetic Waves.* 8th printing. New York, NY: D. Van Nostrad Company.
42. Ni, S., Sun, X., Wang, X., Zhou, G., Yang, F., Wang, J., & He, D. (2010). Low temperature synthesis of Fe_3O_4 micro-spheres and its microwave absorption properties. *Materials Chemistry and Physics, 124*(1), 353–358.
43. Li, Z., Ye, M., Han, A., & Du, H. (2016). Preparation, characterization and microwave absorption properties of $NiFe_2O_4$ and its composites with conductive polymer. *Journal of Materials Science: Materials in Electronics, 27*, 1031–1043.
44. Moitra, D., Hazra, S., Ghosh, B., Jani, R., Patra, M., Vadera, S., & Ghosh, N. (2015). A facile low temperature method for the synthesis of $CoFe_2O_4$ nanoparticles possessing excellent microwave absorption properties. *RSC Advances, 5*(63), 51130–51134.
45. Liu, P., Huang, Y., & Zhang, X. (2015). Cubic $NiFe_2O_4$ particles on graphene–polyaniline and their enhanced microwave absorption properties. *Composites Science and Technology, 107*, 54–60.
46. Li, X., Feng, J., Du, Y., Bai, J., Fan, H., Zhang, H., … Li, F. (2015). One-pot synthesis of $CoFe_2O_4$/graphene oxide hybrids and their conversion into FeCo/graphene hybrids for lightweight and highly efficient microwave absorber. *Journal of Materials Chemistry A, 3*(10), 5535–5546.
47. Khurshid, M. S., Mehmood, M. S., & Zubair, M. (2021). Ultrahigh molecular weight polyethylene and graphene oxide (UHMWPE/GO) nano-composites for EMI shielding. In *Handbook of Polymer and Ceramic Nanotechnology* (pp. 1243–1267). Springer.
48. Srivastava, R. K., Xavier, P., Gupta, S. N., Kar, G. P., Bose, S., & Sood, A. (2016). Excellent electromagnetic interference shielding by graphene-$MnFe_2O_4$-multiwalled carbon nanotube hybrids at very low weight percentage in polymer matrix. *ChemistrySelect, 1*(18), 5995–6003.

49. Anju, Yadav, R. S., Pötschke, P., Pionteck, J. r., Krause, B., Kuřitka, I., . . . Machovsky, M. (2021). High-Performance, lightweight, and flexible thermoplastic polyurethane nanocomposites with Zn^{2+}-substituted $CoFe_2O_4$ nanoparticles and reduced graphene oxide as shielding materials against electromagnetic pollution. *ACS Omega, 6*(42), 28098–28118.
50. Sambyal, P., Dhawan, S., Gairola, P., Chauhan, S. S., & Gairola, S. (2018). Synergistic effect of polypyrrole/BST/RGO/Fe_3O_4 composite for enhanced microwave absorption and EMI shielding in X-Band. *Current Applied Physics, 18*(5), 611–618.
51. Dar, M. A., Majid, K., Najar, M. H., Kotnala, R., Shah, J., Dhawan, S., & Farukh, M. (2017). Surfactant-assisted synthesis of polythiophene/$Ni_{0.5}Zn_{0.5}Fe_{2-x}Ce_xO_4$ ferrite composites: Study of structural, dielectric and magnetic properties for EMI-shielding applications. *Physical Chemistry Chemical Physics, 19*(16), 10629–10643.
52. Pawar, S. P., Gandi, M., & Bose, S. (2016). High performance electromagnetic wave absorbers derived from PC/SAN blends containing multiwall carbon nanotubes and Fe_3O_4 decorated onto graphene oxide sheets. *RSC Advances, 6*(44), 37633–37645.
53. Peng, M., & Qin, F. (2021). Clarification of basic concepts for electromagnetic interference shielding effectiveness. *Journal of Applied Physics, 130*(22), 225108.
54. Aswathi, M., Rane, A. V., Ajitha, A., Thomas, S., & Jaroszewski, M. (2018). EMI shielding fundamentals. *Advanced Materials for Electromagnetic Shielding: Fundamentals, Properties, and Applications*, 1–9.
55. Sankaran, S., Deshmukh, K., Ahamed, M. B., & Pasha, S. K. (2018). Recent advances in electromagnetic interference shielding properties of metal and carbon filler reinforced flexible polymer composites: A review. *Composites Part A: Applied Science and Manufacturing, 114*, 49–71.
56. Pawar, S. P., Biswas, S., Kar, G. P., & Bose, S. (2016). High frequency millimetre wave absorbers derived from polymeric nanocomposites. *Polymer, 84*, 398–419.
57. Chung, D. (2001). Electromagnetic interference shielding effectiveness of carbon materials. *Carbon, 39*(2), 279–285.
58. Shen, Z., Chen, J., Li, B., Li, G., Zhang, Z., & Hou, X. (2020). Recent progress in SiC nanowires as electromagnetic microwaves absorbing materials. *Journal of Alloys and Compounds, 815*, 152388.
59. Rohini, R., & Bose, S. (2017). Electromagnetic wave suppressors derived from crosslinked polymer composites containing functional particles: Potential and key challenges. *Nano-Structures & Nano-Objects, 12*, 130–146.
60. Wang, Y., Du, Y., Xu, P., Qiang, R., & Han, X. (2017). Recent advances in conjugated polymer-based microwave absorbing materials. *Polymers, 9*(1), 29.
61. He, J., Deng, L., Liu, S., Yan, S., Luo, H., Li, Y., . . . Huang, S. (2017). Enhanced microwave absorption properties of Fe_3O_4-modified flaky FeSiAl. *Journal of Magnetism and Magnetic Materials, 444*, 49–53.
62. Yin, X., Kong, L., Zhang, L., Cheng, L., Travitzky, N., & Greil, P. (2014). Electromagnetic properties of Si–C–N based ceramics and composites. *International Materials Reviews, 59*(6), 326–355.
63. Jiang, D., Murugadoss, V., Wang, Y., Lin, J., Ding, T., Wang, Z., . . . Lu, N. (2019). Electromagnetic interference shielding polymers and nanocomposites – a review. *Polymer Reviews, 59*(2), 280–337.
64. Cheng, C., Fan, R., Wang, Z., Shao, Q., Guo, X., Xie, P., . . . Lei, Y. (2017). Tunable and weakly negative permittivity in carbon/silicon nitride composites with different carbonizing temperatures. *Carbon, 125*, 103–112.
65. Quan, B., Liang, X., Ji, G., Cheng, Y., Liu, W., Ma, J., . . . Xu, G. (2017). Dielectric polarization in electromagnetic wave absorption: Review and perspective. *Journal of Alloys and Compounds, 728*, 1065–1075.

66. Chen, Y., Zhang, A., Ding, L., Liu, Y., & Lu, H. (2017). A three-dimensional absorber hybrid with polar oxygen functional groups of MWNTs/graphene with enhanced microwave absorbing properties. *Composites Part B: Engineering, 108*, 386–392.
67. Wang, H., Zhang, Z., Dong, C., Chen, G., Wang, Y., & Guan, H. (2017). Carbon spheres@ MnO_2 core-shell nanocomposites with enhanced dielectric properties for electromagnetic shielding. *Scientific Reports, 7*(1), 15841.
68. Singh, K., Ohlan, A., Pham, V. H., Balasubramaniyan, R., Varshney, S., Jang, J., . . . Dhawan, S. (2013). Nanostructured graphene/Fe_3O_4 incorporated polyaniline as a high performance shield against electromagnetic pollution. *Nanoscale, 5*(6), 2411–2420.
69. Pawar, S. P., Kumar, S., Jain, S., Gandi, M., Chatterjee, K., & Bose, S. (2016). Synergistic interactions between silver decorated graphene and carbon nanotubes yield flexible composites to attenuate electromagnetic radiation. *Nanotechnology, 28*(2), 025201.
70. Zheng, Y., Wang, X., Wei, S., Zhang, B., Yu, M., Zhao, W., & Liu, J. (2017). Fabrication of porous graphene-Fe_3O_4 hybrid composites with outstanding microwave absorption performance. *Composites Part A: Applied Science and Manufacturing, 95*, 237–247.
71. Jian, X., Wu, B., Wei, Y., Dou, S. X., Wang, X., He, W., & Mahmood, N. (2016). Facile synthesis of Fe_3O_4/GCs composites and their enhanced microwave absorption properties. *ACS Applied Materials & Interfaces, 8*(9), 6101–6109.
72. Miles, P., Westphal, W., & Von Hippel, A. (1957). Dielectric spectroscopy of ferromagnetic semiconductors. *Reviews of Modern Physics, 29*(3), 279.
73. Griffiths, J. H. (1946). Anomalous high-frequency resistance of ferromagnetic metals. *Nature, 158*(4019), 670–671.
74. Kong, L. B., Li, Z., Liu, L., Huang, R., Abshinova, M., Yang, Z., . . . Matitsine, S. (2013). Recent progress in some composite materials and structures for specific electromagnetic applications. *International Materials Reviews, 58*(4), 203–259.
75. Singh, A. K., Shishkin, A., Koppel, T., & Gupta, N. (2018). A review of porous lightweight composite materials for electromagnetic interference shielding. *Composites Part B: Engineering, 149*, 188–197.
76. Chung, D. (2020). Materials for electromagnetic interference shielding. *Materials Chemistry and Physics, 255*, 123587.
77. Wang, F., Wang, X., Zhu, J., Yang, H., Kong, X., & Liu, X. (2016). Lightweight $NiFe_2O_4$ with controllable 3D network structure and enhanced microwave absorbing properties. *Scientific Reports, 6*(1), 37892.
78. Aharoni, A. (1991). Exchange resonance modes in a ferromagnetic sphere. *Journal of Applied Physics, 69*(11), 7762–7764.
79. Wen, F., Zhang, F., & Liu, Z. (2011). Investigation on microwave absorption properties for multiwalled carbon nanotubes/Fe/Co/Ni nanopowders as lightweight absorbers. *The Journal of Physical Chemistry C, 115*(29), 14025–14030.
80. Toneguzzo, P., Acher, O., Viau, G., Fiévet-Vincent, F., & Fievet, F. (1997). Observations of exchange resonance modes on submicrometer sized ferromagnetic particles. *Journal of Applied Physics, 81*(8), 5546–5548.
81. Wen, S., Liu, Y., Zhao, X., & Fan, Z. (2015). Synthesis, permeability resonance and microwave absorption of flake-assembled cobalt superstructure. *Journal of Magnetism and Magnetic Materials, 385*, 182–187.
82. Liu, X., Chen, Y., Hao, C., Ye, J., Yu, R., & Huang, D. (2016). Graphene-enhanced microwave absorption properties of Fe_3O_4/SiO_2 nanorods. *Composites Part A: Applied Science and Manufacturing, 89*, 40–46.
83. Zhao, B., Hamidinejad, M., Wang, S., Bai, P., Che, R., Zhang, R., & Park, C. B. (2021). Advances in electromagnetic shielding properties of composite foams. *Journal of Materials Chemistry A, 9*(14), 8896–8949.

84. Munalli, D., Dimitrakis, G., Chronopoulos, D., Greedy, S., & Long, A. (2019). Electromagnetic shielding effectiveness of carbon fibre reinforced composites. *Composites Part B: Engineering, 173*, 106906.
85. Yusoff, A., Abdullah, M., Ahmad, S., Jusoh, S., Mansor, A., & Hamid, S. (2002). Electromagnetic and absorption properties of some microwave absorbers. *Journal of Applied Physics, 92*(2), 876–882.
86. Lv, H., Ji, G., Liang, X., Zhang, H., & Du, Y. (2015). A novel rod-like MnO_2@Fe loading on graphene giving excellent electromagnetic absorption properties. *Journal of Materials Chemistry C, 3*(19), 5056–5064.
87. Ghosh, S., Remanan, S., Mondal, S., Ganguly, S., Das, P., Singha, N., & Das, N. C. (2018). An approach to prepare mechanically robust full IPN strengthened conductive cotton fabric for high strain tolerant electromagnetic interference shielding. *Chemical Engineering Journal, 344*, 138–154.
88. Yang, Z., Zhang, Y., & Wen, B. (2019). Enhanced electromagnetic interference shielding capability in bamboo fiber@ polyaniline composites through microwave reflection cavity design. *Composites Science and Technology, 178*, 41–49.
89. Fan, G., Xiong, T., Mouldi, A., Bouallegue, B., Tran, N., & Mahmoud, M. Z. (2022). Enhanced electromagnetic interference shielding effectiveness of h-BN decorated micro cube-like CaTiO3/Cu nanocomposite. *Ceramics International, 48*(6), 8529–8539.
90. Gill, N., Gupta, V., Tomar, M., Sharma, A. L., Pandey, O., & Singh, D. P. (2020). Improved electromagnetic shielding behaviour of graphene encapsulated polypyrrole-graphene nanocomposite in X-band. *Composites Science and Technology, 192*, 108113.

9 Spinel and Inverse Spinel Ferrites for Environmental Remediation

Harpreet Kaur Channi and Pulkit Kumar

9.1 INTRODUCTION

The creation and use of inverse ferrite and spinel materials for pollution cleanup have recently attracted a lot of attention. Water and air pollution, soil contamination, and hazardous waste management are just a few of the environmental problems that these materials can solve because of their exceptional qualities and adaptability. Ceramics such as spinel and inverse ferrites have the general formula AB_2O_4, where A and B are the names of metal cations that occupy certain crystallographic positions within the spinel structure. Because of their remarkable surface area, chemical stability, magnetic behavior, and catalytic activity, spinel and inverse ferrites are excellent choices for several environmental cleanup tasks. The composition, shape, and surface properties of these materials may be precisely controlled during synthesis utilizing a variety of processes, including sol–gel, hydrothermal, co-precipitation, and combustion synthesis. Several applications of spinel and inverse spinel ferrite materials have been investigated in the field of environmental remediation [1]. These include:

- For water purification, spinel and inverse ferrites are effective adsorbents that can remove organic impurities, heavy metals, and emerging contaminants from water. Their adaptable surface chemistry and large surface area make them ideal for adsorption and the subsequent removal of pollutants via chemical precipitation, surface complexation, and ion exchange.
- Spinel and inverse ferrites are potential catalysts for use in environmental cleanup because of their catalytic activity. They have catalytic applications in wastewater treatment systems, where they may degrade organic contaminants, oxidize toxic gases, and reduce pollutants. Their distinctive electrical and structural features allow them to activate reactant molecules and enable

 DOI: 10.1201/9781032665658-9

moderate catalytic reactions, which is where their catalytic activity comes from [2].

- Recovery and separation from environmental matrices during remediation methods are made easier by the magnetic characteristics of spinel and inverse ferrites. To enable their reuse and minimize secondary contamination, adsorbents based on spinel and inverse spinel ferrite may be recovered from water or soil matrices using magnetic separation methods.
- As photocatalysts, spinel and inverse ferrite materials may degrade organic contaminants when exposed to ultraviolet or visible light. By transferring electrons from the valence to the conduction band through photoexcitation, these materials produce reactive oxygen species (ROS). These species oxidize and break down organic pollutants, resulting in their harmless breakdown into byproducts [3].

Due to their unique characteristics and numerous uses, spinel and inverse spinel ferrite materials show considerable potential for environmental cleanup. Further investigation and improvement of these materials' effectiveness in tackling new environmental problems and advancing sustainable remediation solutions require ongoing research and development in this area. The environmental challenges facing our world today are both complex and pressing. From the contamination of water and soil to the ever-increasing levels of air pollution, humanity finds itself at a critical juncture. The need for innovative, sustainable, and efficient solutions to remediate environmental pollutants has never been greater. In this context, the remarkable properties of spinel and inverse ferrites have emerged as potential game-changers in the field of environmental remediation. This introduction provides a comprehensive overview of the subject, laying the foundation for a deeper exploration of the potential applications and benefits these materials offer in addressing environmental pollution.

There has never been an environmental catastrophe on Earth like this one. Industrial processes, farming, and the combustion of fossil fuels are just a few of the many causes of pollution that have released numerous pollutants into our ecosystems. The environment and human health are greatly endangered by these pollutants, which include organic compounds, heavy metals, and dangerous chemicals, as shown in Figure 9.1. Ecosystem deterioration, animal extinction, and water and soil contamination are only a few of the many negative outcomes of this pollution. Climate change is making these problems worse by amplifying the effects of pollution all across the world. In the face of these difficulties, it is now crystal clear that reducing pollution is an absolute must, not a question of personal preference.

Traditional methods of environmental remediation, while effective to some extent, often fall short of providing sustainable, cost-effective, and comprehensive solutions. These methods include physical removal, chemical treatment, and bioremediation, among others. While they have demonstrated success in specific scenarios, they are often limited by factors such as high costs, slow remediation rates, and the generation of secondary pollutants. The limitations of conventional approaches have fueled the search for innovative technologies that can more effectively and efficiently

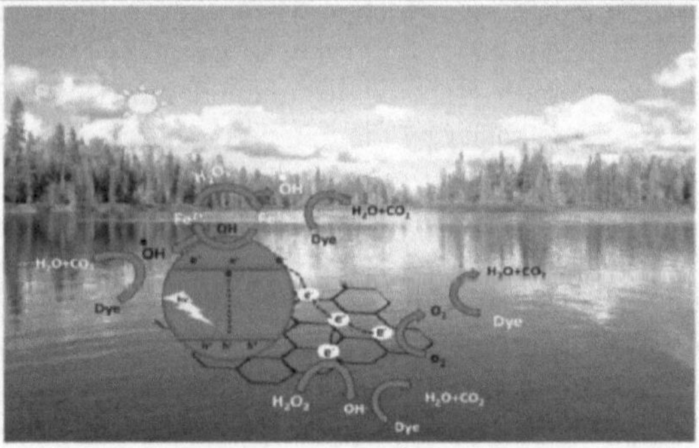

FIGURE 9.1 Spinel ferrite material for environment remediation [4] (CC BY 4.0).

mitigate environmental pollution. This search has led us to explore the potential of spinel and inverse spinel ferrite materials. At the heart of this chapter lies the exploration of spinel and inverse spinel ferrites, two categories of materials that have attracted considerable attention due to their exceptional properties and their potential for addressing environmental contaminants. These materials, characterized by their unique crystal structures and magnetic behavior, offer intriguing opportunities in the realm of environmental remediation.

9.2 LITERATURE REVIEW

Numerous polluted groundwater sites across the globe require an active remediation strategy to restore natural environmental conditions and the local ecosystem. The author used a dynamic weight-enabled Physics-Informed Neural Network (dw-PINN) framework to model the nanoparticle b-type inverse spinel structure with a lattice parameter located at 0.8284 nm and 0.8322 nm [4]. Another researcher studied the biosynthesis, physicochemical, and magnetic properties of the inverse spinel nickel ferrite system. Indicating the development of ferrite material, Fourier transform infrared (FTIR) analysis shows two distinctive vibrational modes around 600 cm^{-1} and 400 cm^{-1} behavior within an aquifer. The X-ray diffraction (XRD) results display the formation of stoichiometric $NiFe_2O_4$ [5]. The author discussed that carbon nanomaterials derived from agricultural waste streams present an exciting material platform that hits multiple sustainability targets by reducing waste entering landfills and enabling clean energy and environmental remediation technologies. The energy and photocatalytic properties of reduced graphene oxide fabricated from coconut coir using a simple reduction method with ferrocene are substantially improved by introducing metallic oxide flakes [6]. Another study provides mercury remediation from wastewater through its spontaneous adsorption on non-functionalized inverse spinel magnetic ferrite nanoparticles. Inverse spinel cubic ferrites MFe_2O_4 (M=Fe and Co) have been fabricated for the high-capacity adsorptive removal of Hg(II) ions. They demonstrated a viable and cost-effective solution for the efficient remediation of Hg ions from wastewater using non-functionalized ferrite adsorbents. Various applications were discussed, such as environmental (treatment of water/wastewater, photocatalytic degradation of dyes, and nano-sorbent for environmental remediation) and biomedical (nano-biosensors for cancer diagnosis at the primary stage, effective targeted drug delivery, magnetic resonance imaging, hyperthermia, and potential drug candidates against cancer and microbial infections) [7]. The author offered comprehensive knowledge on how to choose appropriate natural resources for the green synthesis of nanocobalt ferrite ($CoFe_2O_4$) and the benefits of this approach compared to conventional methods [8]. They presented the results of the synthesis and characterization of nanoparticles of $CoFe_2O_4$ and nickel ferrite ($NiFe_2O_4$) using the co-precipitation method followed by application for the removal of hazardous organic textile dyes of thiazole yellow G (TYG) and alizarin yellow R (AYR). XRD analysis confirmed the formation of cubic spinel structures with average crystallite sizes at 16.07 nm and 13.84 nm for $CoFe_2O_4$ and $NiFe_2O_4$, respectively. Rare earth metal-doping spinel ferrites offer excellent electronic, magnetic, and photocatalytic properties, but they have not been well explored for environmental mitigation [9]. The report presented the facile fabrication of novel $CoNdFe_2O_4$ photocatalysts based on Nd incorporated into $CoFe_2O_4$ for the degradation of Rhodamine B under visible light irradiation. Spinel ferrites are versatile, low-cost, and abundant metal oxides with remarkable electronic and magnetic properties, which find several applications [10]. The report described a cellulose nanofiber-mediated green procedure to prepare controlled highly porous nano-corals comprised of spinel Zn-ferrites. The formic acid (CHO)-sensing properties of flame-made inverse spinel $ZnSn_2O_4$ nanostructures were systematically studied by comparing them with their parent oxides, namely ZnO and SnO_2 [11]. The CHO sensing mechanism with an atomic model was proposed to describe the surface reaction of the inverse spinel $ZnSn_2O_4$ structure to CHO adsorption in comparison with that of the parent oxides [12]. The author discussed that nanosized spinel-substituted ferrites (SFs) with the general formula of MFe_2O_4, where M represents a transition metal, are among a group of magnetic nanomaterials attracting researchers' enormous attention because of their excellent catalytic performance, biomedical applications, and capability for environmental remediation [13]. The author highlighted the synthetic approaches for newly developed SF enzymes and also the structural/experimental factors that affect the kinetics and catalytic mechanisms of enzyme-like reactions. Nanosized spinel-substituted ferrites (SFs) with the general formula of MFe_2O_4 where M represents a transition metal, are among a group of magnetic nanomaterials attracting researchers' enormous attention because of their excellent catalytic performance, biomedical applications, and capability for environmental remediation [14]. Another researcher provided indications of the optimum materials for utilization and insight into their efficiency, reusability, and practicality for the removal of pollutants from wastewater streams. Some of the main material focuses are zero-valent iron, iron oxides, spinel ferrites, and natural and waste-based biosorbents. Other cation transition metals with a 2+ oxidation state can also occupy both tetrahedral and/or octahedral sites in the spinel structure, including $MgFe_2O_4$, a partially inverse spinel, and $ZnFe_2O_4$, a normal spinel [15, 16]. The author provided significant structural insight into the (de) lithiation mechanisms for an intriguing class of iron oxide materials. The report of the synthesis of $Ni_{1-x}Zn_xFe_2O_4$ (x=0, 0.1, 0.3, and 0.5) mixed spinel ferrites via the combustion route [17].

9.3 SPINEL AND INVERSE SPINEL FERRITES IN WASTEWATER TREATMENT

Spinel and inverse spinel ferrites are a class of magnetic materials that have potential applications in wastewater treatment. Spinel ferrites have the general formula MFe_2O_4, where M is a divalent metal ion, and inverse ferrites have the formula FeM_2O_4, where M is a trivalent metal ion. These materials have high surface area, chemical stability, and magnetic properties that make them suitable for adsorbing and removing various contaminants from water. Some examples of spinel and inverse spinel ferrites are magnetite (Fe_3O_4), cobalt ferrite ($CoFe_2O_4$), nickel ferrite ($NiFe_2O_4$), and iron chromite ($FeCr_2O_4$) [2]. There are different methods to synthesize spinel and inverse spinel ferrites, such as co-precipitation, hydrothermal, sol–gel, microwave-assisted, and green synthesis. These methods can produce nanoparticles of different sizes, shapes, and compositions, which can affect their adsorption capacity and efficiency. Some of the factors that influence the adsorption process are pH, temperature, contact time, initial concentration of the pollutant, and dosage of the adsorbent.

Spinel and inverse spinel ferrites can be used to remove various types of pollutants from wastewater, such as heavy metals, dyes, pesticides, pharmaceuticals, and organic compounds. They can also be combined with other materials, such as metal oxides, polymers, carbon-based materials, and other compounds, to form nanocomposites that can enhance their photocatalytic activity and degradation performance [18]. Some of the advantages of using spinel and inverse spinel ferrites for wastewater treatment are their low cost, easy separation by applying a magnetic field, and reusability after regeneration. Spinel and inverse spinel ferrites are magnetic materials that can be used as adsorbents or photocatalysts for removing various pollutants from water. They have high surface area, chemical stability, and magnetic properties that make them suitable for this application. Spinel ferrites have the general formula MFe_2O_4, where M is a divalent metal ion, such as Zn, Co, Ni, Mn, etc. Inverse ferrites have the formula FeM_2O_4, where M is a trivalent metal ion, such as Cr, Al, Ga, etc. These materials have a cubic crystal structure with two types of interstitial sites: tetrahedral (A) and octahedral (B). The distribution of the metal ions in these sites determines the magnetic properties of the ferrites [19].

Spinel and inverse spinel ferrites can be synthesized by different methods, such as co-precipitation, hydrothermal, sol–gel, microwave-assisted, and green synthesis. These methods can produce nanoparticles of different sizes, shapes, and compositions, which can affect their adsorption capacity and efficiency. Some of the factors that influence the adsorption process are pH, temperature, contact time, initial concentration of the pollutant, and dosage of the adsorbent.

Spinel and inverse spinel ferrites can be used to remove various types of pollutants from wastewater, such as heavy metals (e.g., Pb, Cd, Cu, Zn, Ni), dyes (e.g., Methylene Blue, Rhodamine B, Malachite Green), pesticides (e.g., atrazine, 2,4-D),

pharmaceuticals (e.g., ibuprofen, diclofenac), and organic compounds (e.g., phenol, benzene). They can also be combined with other materials, such as metal oxides (e.g., TiO_2, ZnO), polymers (e.g., chitosan), carbon-based materials (e.g., graphene oxide), and other compounds (e.g., zeolite), to form nanocomposites that can enhance their photocatalytic activity and degradation performance. Spinel and inverse spinel ferrites have some advantages over other nanomaterials for wastewater treatment. They are low cost, easy to separate by applying a magnetic field, and reusable after regeneration. However, they also have some challenges, such as agglomeration, leaching of metal ions, and low visible light utilization [20].

9.3.1 Adsorption Mechanisms

Adsorption mechanisms for spinel and inverse spinel ferrites in wastewater treatment are the processes by which these magnetic materials interact with the pollutants and remove them from the water. There are different types of adsorption mechanisms, depending on the nature of the adsorbent and the adsorbate, the pH, the temperature, and other factors. Some of the common adsorption mechanisms are as follows:

- **Ion exchange:** This is the exchange of ions between the adsorbent and the adsorbate, resulting in a change in the charge balance of both. For example, some spinel ferrites can exchange their divalent metal ions (M) with heavy metal ions (M) in water, forming MFe_2O_4 and releasing M^{2+} ions into the solution. This mechanism is influenced by the affinity of the metal ions for the ferrite lattice and the concentration gradient of the ions.
- **Complexation:** This is the formation of a coordination bond between the adsorbent and the adsorbate, involving electron donation and acceptance. For example, some inverse ferrites can form complexes with organic molecules, such as dyes, by donating electrons from their iron atoms to the vacant orbitals of the molecules. This mechanism is influenced by the availability of coordination sites on the ferrite surface and the molecular structure of the organic compounds [21].
- **Precipitation:** This is the formation of insoluble compounds on the adsorbent surface, resulting from a chemical reaction between the adsorbent and the adsorbate. For example, some spinel ferrites can react with phosphate ions in water, forming insoluble metal phosphates (MPO_4) on their surface. This mechanism is influenced by the solubility product of the precipitate and the pH of the solution.
- **Lewis acid-base interaction:** This is the interaction between a Lewis acid (electron acceptor) and a Lewis base (electron donor), resulting in a sharing of electron pairs. For example, some spinel ferrites can act as Lewis acids, accepting electrons from Lewis bases, such as hydroxyl groups or carboxyl groups, on the surface of alginate-based adsorbents. This mechanism is influenced by the acidity or basicity of the adsorbent and the adsorbate [22].
- **Electrostatic interaction:** This is the attraction or repulsion between charged particles, resulting from a difference in electric potential. For

example, some spinel ferrites can have a positive or negative surface charge, depending on the pH of the solution, attracting or repelling oppositely or similarly charged pollutants, such as metal ions or anionic dyes. This mechanism is influenced by the surface charge density of the adsorbent and the ionic strength of the solution [23].

9.3.2 Heavy Metal Removal

Heavy metal removal from wastewater using spinel and inverse spinel ferrites is a process of using magnetic materials to adsorb and remove toxic metals from water sources. Spinel and inverse spinel ferrites have the general formula MFe_2O_4 and FeM_2O_4, respectively, where M is a metal ion. These materials have high surface area, chemical stability, and magnetic properties that make them suitable for this application. Spinel and inverse spinel ferrites can be synthesized by different methods, such as co-precipitation, sol–gel, spinning, chemical vapor deposition, pyrolysis, and sputtering. These methods can produce nanoparticles of different sizes, shapes, and compositions, which can affect their adsorption capacity and efficiency. Some of the factors that influence the adsorption process are pH, temperature, contact time, initial concentration of the pollutant, and dosage of the adsorbent [24].

Spinel and inverse spinel ferrites can remove various types of heavy metals from wastewater, such as lead, mercury, cadmium, chromium, copper, nickel, and zinc. The mechanism of removal involves the transport of the heavy metal ions to the surface of the adsorbent particles due to the electrostatic force of attraction between charged particles. Specific surface area, pH, adsorbate concentrations, and charges on the adsorbate/adsorbent are the primary variables involved in the process. Some examples of spinel and inverse spinel ferrites used for heavy metal removal are zinc/manganese-doped nickel ferrites ($Ni_{1-x}M_xFe_2O_4$) where M = Zn/Mn, which can remove Methylene Blue dye and heavy metals under natural solar light illumination and magnetite (Fe_3O_4), which can remove arsenic and chromium from water. Spinel and inverse ferrites have some advantages over other nanomaterials for wastewater treatment. They are low cost, easy to separate by applying a magnetic field, and reusable after regeneration. However, they also have some challenges, such as agglomeration, leaching of metal ions, and low visible light utilization [25].

9.3.3 Surface Modifications and Enhancements

Surface modifications and enhancements for wastewater treatment using spinel and inverse spinel ferrites are methods of improving the performance and efficiency of these magnetic materials for removing pollutants from water sources. Surface modifications involve changing the chemical or physical properties of the surface of the ferrites, such as coating, doping, functionalizing, or grafting with other materials. Enhancements involve increasing the surface area, porosity, or stability of the ferrites, such as reducing particle size, creating hollow structures, or forming nanocomposites [26]. Some examples of surface modifications and enhancements for wastewater treatment using spinel and inverse spinel ferrites are the coating of spinel ferrites with metal oxides, such as TiO_2, ZnO, or CeO_2, to enhance their photocatalytic activity and degradation performance under visible or UV light. The metal oxides can act as electron acceptors or donors, facilitating the charge transfer and separation between the ferrites and the pollutants and doping spinel ferrites with transition metals, such as Co, Ni, Mn, or Cu, to improve their adsorption capacity and magnetic properties. Transition metals can modify the electronic structure and surface charge of the ferrites, increasing their affinity and selectivity for heavy metal ions or organic molecules. Functionalizing spinel ferrites with organic groups, such as carboxyl, amine, or thiol, increases their hydrophilicity and biocompatibility. The organic groups can provide specific binding sites for pollutants, enhancing their removal efficiency and specificity, and grafting spinel ferrites with polymers, such as chitosan, alginate, or polyvinyl alcohol, to improve their stability and reusability. The polymers can prevent the agglomeration and leaching of the ferrites, as well as provide additional adsorption sites for pollutants [7].

Reducing the particle size of spinel ferrites to nanoscale increases their surface area and reactivity. Nanosized ferrites can have higher adsorption sites and faster diffusion rates than bulk ferrites, improving their removal kinetics and equilibrium. Creating hollow structures of spinel ferrites increases their porosity and surface area. Hollow ferrites can have lower density and higher loading capacity than solid ferrites, enhancing their adsorption performance and separation efficiency and forming nanocomposites of spinel ferrites with other materials, such as carbon-based materials (e.g., graphene oxide), zeolite, or other compounds. Nanocomposites can combine the advantages of different materials, such as high surface area, magnetic properties, photocatalytic activity, and adsorption capacity [27].

9.3.4 Magnetic Separation for Efficient Recovery

Magnetic separation for efficient recovery in wastewater treatment using spinel and inverse spinel ferrites is a process of using a magnetic field to separate and recover these magnetic materials after they have adsorbed and removed pollutants from water sources. Magnetic separation is a widely used method because of its low cost, high efficiency, and ease of operation. Some examples of magnetic separation devices are magnetic filters, magnetic rods, magnetic drums, and magnetic nanoparticles. Magnetic separation for wastewater treatment using spinel and inverse spinel ferrites has several advantages over other methods. It can reduce the amount of sludge generated, avoid secondary pollution, and enhance the reusability of the adsorbents [28].

It can also improve the removal efficiency of some pollutants that are difficult to degrade by other methods, such as dyes, pharmaceuticals, and organic compounds. Magnetic separation for wastewater treatment using spinel and inverse spinel ferrites also has some challenges and limitations. It requires a strong magnetic field to separate the ferrites from the water, which may consume a lot of energy and affect the stability of the ferrites. It may also cause some loss of the ferrites during the separation process, which may reduce their adsorption capacity and increase their cost. Moreover, it may not be effective for some pollutants that have a low affinity or weak interaction with the ferrites [29].

9.3.5 Photocatalytic and Fenton-Like Activities

Photocatalytic and Fenton-like activities are two types of catalytic processes that can be used to degrade organic pollutants in water and soil. Photocatalytic activity refers to the ability of a material to generate reactive oxygen species (ROS) such as hydroxyl radicals, superoxide anions, and hydrogen peroxide under light irradiation. These ROS can then oxidize and break down the organic molecules into harmless products such as water and carbon dioxide. Fenton-like activity refers to the ability of a material to catalyze the decomposition of hydrogen peroxide into hydroxyl radicals in the presence of iron ions. These hydroxyl radicals are also very effective in destroying organic contaminants [30].

Spinel and inverse spinel ferrites are two types of metal oxide nanomaterials that have been widely studied for their photocatalytic and Fenton-like activities. Spinel ferrites have the general formula AB_2O_4, where A and B are metal cations, and O is oxygen. Inverse ferrites have the general formula B_2AO_4, where A and B are reversed. The most common spinel and inverse spinel ferrites are based on iron oxide (Fe_3O_4) and other transition metals such as cobalt, nickel, manganese, zinc, copper, etc. These materials have unique magnetic, optical, and electrical properties that make them suitable for various applications [31].

The photocatalytic and Fenton-like activities of spinel and inverse spinel ferrites depend on several factors, such as their composition, structure, morphology, surface area, crystallinity, defect density, doping level, etc. These factors can be controlled by using different synthesis methods, such as sol–gel, hydrothermal, co-precipitation, thermal decomposition, etc. By tuning these factors, the spinel and inverse spinel ferrites can be optimized for different types of organic pollutants and environmental conditions. Some examples of the photocatalytic and Fenton-like activities of spinel and inverse spinel ferrites are as follows:

- **[$ZnFe_2O_4$]** nanoparticles synthesized by the sol–gel method showed high photocatalytic activity for the degradation of Methylene Blue dye under visible light irradiation.
- **[$CoFe_2O_4$]** nanorods synthesized by the hydrothermal method showed enhanced Fenton-like activity for the degradation of phenol in aqueous solution under acidic conditions.
- **[$NiFe_2O_4$]** nanospheres synthesized by the thermal decomposition method showed both photocatalytic and Fenton-like activities for the degradation of Rhodamine B dye under UV light irradiation and hydrogen peroxide addition.
- **[$CuFe_2O_4$]** nanocubes synthesized by the co-precipitation method showed high photocatalytic activity for the degradation of 4-chlorophenol under visible light irradiation [32].

The photocatalytic degradation performance of spinel ferrites and inverse spinel ferrites for various medicines and dyes is shown in Table 9.1. Spinel ferrites degraded medications including Ibuprofen (70%), Paracetamol (75%), and Caffeine (65%), in

TABLE 9.1
Photocatalytic Activities of Spinel and Inverse Spinel Ferrites

Material	Dye Type	Dye Degradation (Percentage Removal)	Drug Type	Drug Degradation (Percentage Removal)
Spinel ferrites	Rhodamine B	85%	Ibuprofen	70%
	Methylene Blue	90%	Paracetamol	75%
	Congo Red	80%	Caffeine	65%
Inverse spinel ferrites	Rhodamine B	95%	Ibuprofen	75%
	Methylene Blue	92%	Paracetamol	80%
	Congo Red	88%	Caffeine	70%

addition to dyes like Rhodamine B (85%), Methylene Blue (90%), and Congo Red (80%). In contrast, inverse spinels showed even more efficacy in degrading pollutants, with remarkable degradation rates of 95% for Rhodamine B, 92% for Methylene Blue, and 88% for Congo Red dyes. Degradation rates of 75% for Ibuprofen, 80% for Paracetamol, and 70% for Caffeine were achieved using inverse spinel, demonstrating better performance in the drug degradation task. Both spinel and inverse spinel ferrites are promising photocatalysts for use in environmental cleanup, according to these findings. Inverse spinel ferrites are great candidates for future research and development in environmental remediation technologies due to their higher degradation rates. This suggests that their unique crystal structure and composition may offer enhanced photocatalytic activity compared to traditional spinel.

9.3.5.1 Photocatalytic Mechanism

The photocatalytic mechanism involves a series of chemical reactions facilitated by the absorption of light energy by a photocatalyst material. This process is commonly employed in various applications, including environmental remediation, water purification, and self-cleaning surfaces. The key steps in the photocatalytic mechanism typically include the following:

- **Absorption of photons:** The photocatalyst material absorbs photons (light particles) from a light source, usually ultraviolet (UV) or visible light.
- **Generation of electron–hole pairs:** Upon absorbing photons, the photocatalyst undergoes electronic excitation, leading to the generation of electron–hole pairs. This process occurs due to the promotion of electrons from the valence band to the conduction band, leaving behind positively charged holes in the valence band.
- **Redox reactions:** The photogenerated electrons and holes participate in redox (oxidation–reduction) reactions with adsorbed species on the photocatalyst surface or in the surrounding environment. The electrons can reduce oxidizing species, while the holes can oxidize reducing species.

- **Generation of reactive oxygen species (ROS):** One of the primary outcomes of the redox reactions is the formation of reactive oxygen species (ROS) such as hydroxyl radicals (•OH), superoxide radicals ($\bullet O^{2-}$), and hydrogen peroxide (H_2O_2). These ROS are highly reactive and play a crucial role in the degradation of organic pollutants or microbial cells.
- **Degradation or transformation of target molecules:** The generated ROS attack organic pollutants, pathogens, or other target molecules adsorbed on the photocatalyst surface or present in the surrounding medium. This leads to the degradation of pollutants into harmless byproducts or the sterilization of microbial cells.
- **Regeneration of photocatalysts:** After participating in the redox reactions, the photogenerated electrons and holes may recombine or react with other species. To sustain the photocatalytic activity, the photocatalyst needs to maintain a continuous supply of light energy to generate new electron–hole pairs. Overall, the photocatalytic mechanism harnesses light energy to drive chemical transformations, offering a sustainable and environmentally friendly approach for various applications requiring pollutant degradation, disinfection, or self-cleaning properties as shown in Figure 9.2.

9.3.5.2 ROS Generation and Its Role

Reactive oxygen species (ROS) are molecules that contain oxygen and have one or more unpaired electrons, making them highly reactive and unstable. ROS can be generated by various sources in the cell, such as the mitochondria, the endoplasmic reticulum, the plasma membrane, and the peroxisomes. ROS can also be induced by external factors, such as UV radiation, chemicals, drugs, and pathogens. ROS have a

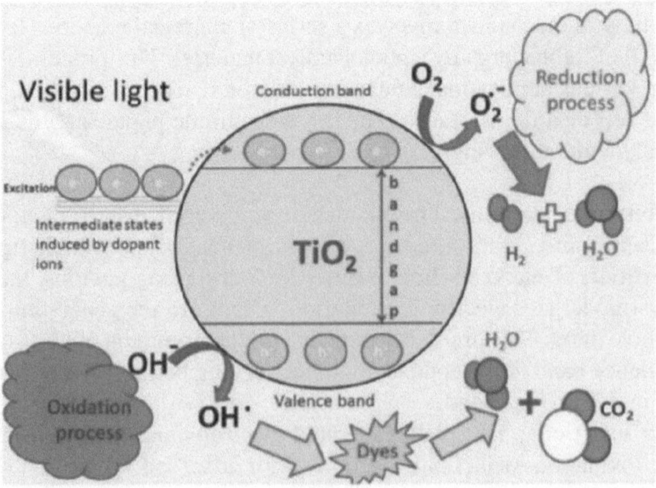

FIGURE 9.2 Photocatalytic mechanism of $CoFe_2O_4$ [33] (CC BY 4.0).

dual role in the cell: they can act as signaling molecules that regulate various cellular processes, such as cell growth, differentiation, migration, and apoptosis; or they can cause oxidative damage to biomolecules, such as DNA, proteins, and lipids, leading to cell dysfunction and death. Therefore, the level of ROS in the cell must be tightly controlled by a balance between ROS generation and ROS scavenging [34].

The cell has various antioxidant defense systems that can scavenge ROS and prevent oxidative stress. These systems include enzymatic antioxidants, such as superoxide dismutase (SOD), catalase (CAT), glutathione peroxidase (GPX), and peroxiredoxin (PRX); and non-enzymatic antioxidants, such as glutathione (GSH), vitamin C, vitamin E, and carotenoids. These antioxidants can either directly neutralize ROS or repair the damage caused by ROS [35]. ROS generation and its role in the cell are influenced by many factors, such as the type of ROS, the source of ROS, the location of ROS production, the intensity and duration of ROS exposure, the availability of antioxidants, and the type of cell. Different types of cells may have different sensitivities and responses to ROS. For example, some cells may use ROS as a defense mechanism against pathogens or tumor cells, while other cells may be more vulnerable to ROS-induced damage or apoptosis [36].

9.3.6 Organic Pollutant Degradation

Organic pollutant degradation is the process of breaking down organic compounds that are harmful to the environment and human health. Organic pollutants can be classified into two categories: biodegradable and non-biodegradable. Biodegradable organic pollutants can be decomposed by microorganisms or natural processes, such as sunlight, water, and oxygen. Non-biodegradable organic pollutants, also known as persistent organic pollutants (POPs), are resistant to degradation and can persist in the environment for a long time, accumulating in the food chain and causing adverse effects on wildlife and humans. There are various methods to degrade organic pollutants, such as chemical, physical, and biological approaches. Chemical methods involve the use of chemicals, such as oxidants, reductants, acids, bases, or catalysts, to transform organic pollutants into less toxic or more soluble substances. Physical methods involve the use of physical forces, such as heat, light, electricity, ultrasound, or radiation, to destroy the molecular structure of the organic pollutants. Biological methods involve the use of living organisms, such as bacteria, fungi, algae, or plants, to metabolize organic pollutants into harmless products [37].

Among these methods, photocatalysis is a promising technique that uses light energy to activate a photocatalyst, which is a material that can generate reactive oxygen species (ROS) under light irradiation. ROS are highly reactive molecules that can oxidize and degrade organic pollutants into carbon dioxide, water, and inorganic ions. Photocatalysis has several advantages over other methods, such as low cost, high efficiency, environmental friendliness, and wide applicability. Many types of photocatalysts have been studied for organic pollutant degradation, such as metal oxides (e.g., TiO_2, ZnO, Fe_2O_3), metal sulfides (e.g., CdS, ZnS), metal-organic frameworks (MOFs), carbon-based materials (e.g., graphene, carbon nanotubes), and hybrid materials (e.g., metal oxide/carbon composites). These photocatalysts can be modified by doping with other elements, coupling with other materials, or creating defects to enhance their photocatalytic performance [38, 39].

The photocatalytic activity of these materials depends on various factors, such as their composition, structure, morphology, surface area, crystallinity, defect density, doping level, etc. The photocatalytic degradation of organic pollutants can be influenced by several parameters, such as the type and concentration of the pollutant, the type and amount of the photocatalyst, the light source and intensity, the pH and temperature of the solution, and the presence of oxygen and other additives (e.g., hydrogen peroxide).These parameters need to be optimized to achieve the best degradation efficiency and selectivity [40].

9.4 ENVIRONMENTAL FATE OF FERRITE NANOPARTICLES

The environmental fate of ferrite nanoparticles is a topic that concerns the potential impacts of these nanomaterials on the environment and human health. Ferrite nanoparticles are magnetic particles that consist of iron oxide and other metal oxides, such as cobalt, zinc, copper, nickel, manganese, etc. They have various applications in fields such as electronics, biomedicine, catalysis, and environmental remediation. However, these nanoparticles can also pose some risks to the environment and human health due to their unique properties and behavior. For example, they can generate reactive oxygen species (ROS) that can cause oxidative stress and damage to cells and biomolecules. They can also interact with other environmental factors, such as pH, salinity, organic matter, and microorganisms, that can affect their stability, aggregation, dissolution, transformation, and bioavailability. These processes can influence their transport, distribution, accumulation, and toxicity in different environmental compartments, such as water, soil, sediment, and biota [41].

Therefore, it is important to understand the environmental fate of ferrite nanoparticles and their potential effects on the ecosystem and human health. Various methods and techniques can be used to study the environmental fate of ferrite nanoparticles, such as analytical methods (e.g., spectroscopy, microscopy), modeling methods (e.g., mass balance models, fate models), and experimental methods (e.g., laboratory tests, field studies). These methods can provide useful information on the physicochemical properties, environmental behavior, and biological interactions of ferrite nanoparticles. Various strategies can be used to reduce the environmental risks of ferrite nanoparticles, such as green synthesis methods (e.g., using natural or biodegradable materials), surface modification methods (e.g., using coatings or functional groups), and safe disposal methods (e.g., using filters or adsorbents). These strategies can help improve the biocompatibility, stability, and recyclability of ferrite nanoparticles [42].

9.4.1 Ecological Implications and Risk Assessment

Ferrite nanoparticles, tiny particles composed of iron oxides with unique magnetic and electrical properties, have gained widespread use in various industrial and technological applications. However, their introduction into the environment raises concerns about potential ecological implications and the need for comprehensive risk assessments. In this article, we will delve into the ecological implications of ferrite nanoparticles and explore the essential elements of risk assessment associated with these nanomaterials.

9.4.1.1 Ecological Implications

- **Toxicity:** One of the most significant ecological concerns regarding ferrite nanoparticles is their potential toxicity to various organisms. These nanoparticles can release iron ions, which, in elevated concentrations, can be harmful to living organisms. Aquatic organisms, such as fish and invertebrates, are particularly vulnerable to these nanoparticles when they enter water bodies. Iron ions can disrupt cellular processes, leading to adverse effects on growth, reproduction, and overall health. In soil environments, soil-dwelling organisms, including earthworms and microorganisms, may also be impacted by the presence of ferrite nanoparticles. Toxicity can disrupt soil ecosystems, affecting nutrient cycling and plant growth [43].
- **Bioaccumulation:** Ferrite nanoparticles, like many nanoparticles, have the potential to bioaccumulate in the food chain. When smaller organisms in aquatic ecosystems, such as algae and zooplankton, ingest these nanoparticles, they can be transferred up the food chain as larger organisms, like fish, consume the smaller ones. As a result, there is a risk of biomagnification, where the concentration of nanoparticles increases at higher trophic levels. This can lead to harmful effects on top predators, including birds or mammals that rely on contaminated prey. Bioaccumulation of nanoparticles can disrupt the balance of ecosystems and impact biodiversity.
- **Water and soil contamination:** The release of ferrite nanoparticles into water bodies or soil can result in contamination and alter the quality of these ecosystems. These nanoparticles may enter the environment through various routes, including industrial discharges or the use of nanomaterial-containing products. Once released, they can disperse in water or soil, affecting local flora and fauna. Water bodies can experience reduced water quality, potentially leading to oxygen depletion and the decline of aquatic life. Soil contamination can disrupt nutrient cycling, impair plant growth, and affect the entire soil ecosystem.
- **Altered microbial communities:** Ferrite nanoparticles can influence microbial communities in soil. Microbes play a crucial role in maintaining soil health and ecosystem functioning by decomposing organic matter, fixing nitrogen, and aiding in nutrient cycling. The presence of nanoparticles can alter the composition and activity of these communities. Some studies suggest that certain nanoparticles can inhibit the growth of beneficial soil microbes, potentially leading to shifts in microbial diversity and reduced nutrient availability. These changes in soil microbial communities can have cascading effects on plant health and the overall ecosystem [44].

9.4.1.2 Risk Assessment

A robust risk assessment process is vital for addressing the ecological implications of ferrite nanoparticles. This process comprises several key components. Firstly, hazard identification involves recognizing potential hazards linked to ferrite nanoparticles, encompassing their physical and chemical properties, toxicological effects, and likelihood of being released into the environment. Researchers rely on existing data and experimental studies to identify ecological risks. Exposure assessment follows, where the evaluation of how organisms in the environment might come into contact with these nanoparticles is examined. It considers pathways of release and transport, as well as potential routes of exposure for various species [45].

Toxicity assessment aims to determine the adverse effects of ferrite nanoparticles on different organisms, conducting laboratory experiments to understand the mechanisms by which these nanoparticles may harm living organisms. Subsequently, risk characterization combines information from hazard identification, exposure assessment, and toxicity assessment to estimate the potential ecological risks posed by ferrite nanoparticles, quantifying the likelihood and magnitude of harm and identifying vulnerable species and ecosystems. Risk management, based on the findings of the risk assessment, involves developing strategies to manage and mitigate identified risks, which may encompass regulatory measures, guidelines for safe handling and disposal, and recommendations for the environmentally responsible use of ferrite nanoparticles. Lastly, continuous monitoring and research are crucial for comprehending the long-term effects of ferrite nanoparticles in the environment, refining risk assessments, detecting emerging issues, and informing adaptive management strategies [46].

9.4.1.3 Risk Mitigation and Regulations

To mitigate the ecological risks linked to ferrite nanoparticles, a combination of regulatory measures and responsible practices is crucial. Various strategies can be employed for this purpose. Firstly, environmental regulations can be established by governments and regulatory agencies, setting guidelines for the use and disposal of ferrite nanoparticles. These regulations may include restrictions on nanoparticle concentrations in industrial effluents and requirements for product labeling to inform consumers about potential risks associated with these materials. Furthermore, it is advisable for companies and organizations engaged in the production or use of ferrite nanoparticles to conduct comprehensive environmental impact assessments [47].

These assessments help identify and address potential risks before introducing nanoparticles into the environment, ensuring that precautionary measures are in place. Safe handling and disposal practices are equally essential, with workers following strict safety protocols to prevent accidental releases, and environmentally sound disposal methods should be established to minimize contamination. Additionally, ongoing research into the environmental behavior of ferrite nanoparticles and the development of safer alternatives within the realm of materials science can contribute to the reduction of ecological risks. Such innovations can lead to the design of nanoparticles with reduced toxicity and improved biocompatibility. Lastly, public awareness and education campaigns are valuable tools to inform consumers, businesses, and the general public about the potential risks associated with ferrite nanoparticles. These efforts promote responsible usage and disposal, fostering a more environmentally conscious approach to the utilization of these nanomaterials.

9.5 EMERGING TRENDS AND FUTURE PROSPECTS

In the realm of ferrite nanoparticles, emerging trends and prospects are significantly influencing the landscape of various fields. One notable trend is the development of advanced ferrite-based composites. These composites combine ferrite nanoparticles with other materials, creating versatile and multifunctional substances. By doing so, these materials become more adaptable and serve various functions. For instance, they can be used in data storage, sensors, and even medical devices, owing to their flexible and magnetic properties. Moreover, these advanced composites offer improved mechanical, electrical, and magnetic characteristics, making them suitable for applications in electronics, aerospace, and automotive industries [48].

Researchers can tailor these composites to meet specific needs, enabling applications in electromagnetic interference shielding, drug delivery systems, and environmental remediation. Another promising avenue is the integration of ferrite nanoparticles into state-of-the-art environmental remediation technologies. The magnetic properties of ferrite nanoparticles make them ideal for the removal of pollutants from water and soil. They can be employed in magnetic adsorbents and magnetic nanoparticle systems that efficiently eliminate heavy metals, organic contaminants, and even radioactive substances. Furthermore, in-situ remediation techniques using ferrite nanoparticles enable the treatment of contaminated sites without extensive excavation, minimizing costs and ecosystem disruption. This integration also extends to wastewater treatment plants and systems, enhancing pollutant removal, and thereby improving water quality [49].

Sensors utilizing ferrite nanoparticles offer advanced real-time monitoring of contaminant presence, enabling more efficient and timely remediation efforts. Additionally, for sustainable implementation and regulatory aspects, several key considerations come into play. It is essential to conduct thorough environmental impact assessments to understand potential risks and develop strategies for mitigation [50]. Regulatory frameworks are required at both national and international levels to address the production, use, and disposal of ferrite nanoparticles. These frameworks may include guidelines for safe handling, exposure limits, and labeling requirements. Moreover, sustainable implementation calls for the development of recycling and disposal methods to reduce the potential for environmental contamination. Public awareness campaigns that educate the public about the benefits and potential risks of ferrite nanoparticles can foster responsible use and disposal, encouraging environmentally conscious choices as shown in Figure 9.3.

In an era of international collaboration, it is crucial to establish common safety standards and ensure the responsible implementation of ferrite nanoparticles. These trends collectively signify the potential for improved materials and environmentally responsible practices in various sectors, spanning electronics, healthcare,

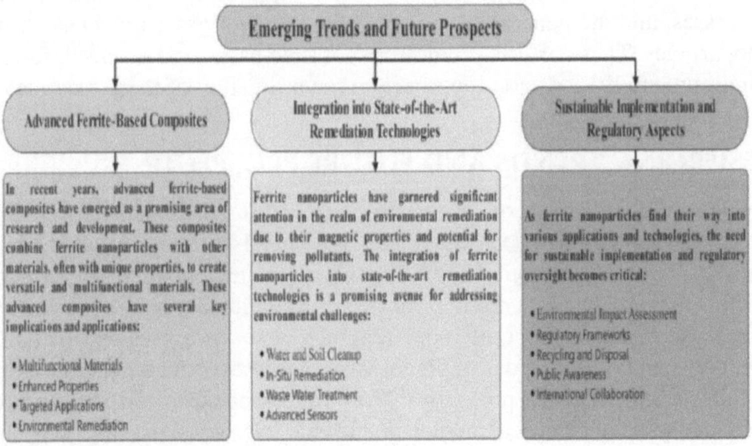

FIGURE 9.3 Emerging trends and prospects for ferrite nanoparticles.

environmental protection, and more. As technology and our understanding of materials continue to advance, ferrite nanoparticles will undoubtedly play a pivotal role in shaping a more versatile, efficient, and sustainable future [51].

9.6 CONCLUSION

The research discussed in this chapter showcases the versatility and potential of spinel and inverse spinel ferrites in various environmental remediation applications. Their unique crystal structures, magnetic properties, and adsorption mechanisms make them valuable candidates for heavy metal removal and wastewater treatment. Furthermore, their photocatalytic and Fenton-like activities, coupled with the generation of reactive oxygen species, demonstrate their efficacy in organic pollutant degradation and the potential for synergistic utilization with renewable energy sources. However, in our pursuit of environmental remediation, it is imperative to consider the broader context of environmental safety and sustainability. Understanding the environmental fate of ferrite nanoparticles, assessing ecological implications, and making responsible choices regarding their use and disposal are critical aspects that deserve careful attention. Looking ahead, emerging trends and prospects in this field point to the development of advanced ferrite-based composites, integration into state-of-the-art remediation technologies, and the continued emphasis on sustainable implementation while navigating regulatory aspects. In summary, the investigation of spinel and inverse spinel ferrites in environmental remediation offers not only a valuable understanding of their applications but also underscores the importance of responsible and sustainable practices in the quest for cleaner and safer environments. As we continue to innovate and adapt, these materials hold the promise of contributing significantly to the preservation and restoration of our natural surroundings.

REFERENCES

1. Park, C.M., et al., Potential utility of graphene-based nano spinel ferrites as adsorbent and photocatalyst for removing organic/inorganic contaminants from aqueous solutions: A mini-review. *Chemosphere*, 2019. **221**: p. 392–402.
2. Kefeni, K.K., B.B. Mamba, and T.A. Msagati, Application of spinel ferrite nanoparticles in water and wastewater treatment: A review. *Separation and Purification Technology*, 2017. **188**: p. 399–422.
3. Kaur, M., and N. Kaur, Ferrites: Synthesis and applications for environmental remediation, in *Ferrites and Ferrates: Chemistry and Applications in Sustainable Energy and Environmental Remediation*. 2016, ACS Publications. p. 113–136.
4. Nilabh, S., and F. Grandia, Dynamic weights enabled physics-informed neural network for simulating the mobility of engineered Nano-particles in a contaminated aquifer. arXiv preprint arXiv:2211.03525, 2022.
5. Al-Senani, G.M., et al., Biosynthesis, physicochemical and magnetic properties of inverse spinel nickel ferrite system. *Crystals*, 2022. **12**(11): p. 1542.
6. Ramasamy, T., et al., Spinel CoFe2O4 Nanoflakes: A path to enhance energy generation and environmental remediation potential of waste-derived rGO. *Nanomaterials*, 2022. **12**(21): p. 3822.
7. Viltres, H., et al., Mercury remediation from wastewater through its spontaneous adsorption on non-functionalized inverse spinel magnetic ferrite nanoparticles. *Environmental Technology*, 2022. **45**(6): p. 1155–1168.
8. Tamboli, Q.Y., et al., Green synthesis of cobalt ferrite nanoparticles: An emerging material for environmental and biomedical applications. *Journal of Nanomaterials*, 2023. **1**: p. 9770212.
9. Patil, D.J., and S.N. Behera, Synthesis and characterization of nanoparticles of cobalt and nickel ferrites for elimination of hazardous organic dyes from industrial wastewater. *Environmental Science and Pollution Research*, 2023. **30**(18): p. 53323–53338.
10. Nguyen, L.T., et al., Efficient and recyclable Nd 3+-doped CoFe 2 O 4 for boosted visible light-driven photocatalytic degradation of Rhodamine B dye. *RSC Advances*, 2023. **13**(16): p. 10650–10656.
11. Teixeira, L.T., et al., Sustainable cellulose nanofibers-mediated synthesis of uniform spinel Zn-ferrites nanocorals for high performances in supercapacitors. *International Journal of Molecular Sciences*, 2023. **24**(11): p. 9169.
12. Punginsang, M., et al., Comparative study on formic acid sensing properties of flame-made Zn_2SnO_4 nanoparticles and its parent metal oxides. *Physical Chemistry Chemical Physics*, 2023. **25**(22): p. 15407–15421.
13. Saleh, T.S., et al., Design and development of novel composites containing nickel ferrites supported on activated carbon derived from agricultural wastes and its application in water remediation. *Materials*, 2023. **16**(6): p. 2170.
14. Chaibakhsh, N., and Z. Moradi-Shoeili, Enzyme mimetic activities of spinel substituted nanoferrites (MFe_2O_4): A review of synthesis, mechanism and potential applications. *Materials Science and Engineering: C*, 2019. **99**: p. 1424–1447.
15. Maksoud, M.A., et al., Insight on water remediation application using magnetic nanomaterials and biosorbents. *Coordination Chemistry Reviews*, 2020. **403**: p. 213096.
16. Bock, D.C., et al., (De) lithiation of spinel ferrites Fe_3O_4, $MgFe_2O_4$, and $ZnFe_2O_4$: a combined spectroscopic, diffraction and theory study. *Physical Chemistry Chemical Physics*, 2020. **22**(45): p. 26200–26215.
17. Dhiman, P., et al., Nanostructured magnetic inverse spinel Ni–Zn ferrite as environmentally friendly visible light driven photo-degradation of levofloxacin. *Chemical Engineering Research and Design*, 2021. **175**: p. 85–101.

18. Venturini, J., Application of ferrite nanoparticles in wastewater treatment. *Nanomaterials for Eco-friendly Applications*, 2019: p. 141–151.
19. Suresh, R., et al., Recent advancements of spinel ferrite based binary nanocomposite photocatalysts in wastewater treatment. *Chemosphere*, 2021. **274**: p. 129734.
20. Lingamdinne, L.P., et al., Preparation and characterization of porous reduced graphene oxide based inverse spinel nickel ferrite nanocomposite for adsorption removal of radionuclides. *Journal of Hazardous Materials*, 2017. **326**: p. 145–156.
21. Kefeni, K.K., and B.B. Mamba, Photocatalytic application of spinel ferrite nanoparticles and nanocomposites in wastewater treatment. *Sustainable Materials and Technologies*, 2020. **23**: p. e00140.
22. Sonia, et al., Spinel ferrites/metal oxide nanocomposites for wastewater treatment. *Applied Physics A*, 2023. **129**(2): p. 91.
23. Narayana, P., et al., Predictive capability evaluation and optimization of Pb (II) removal by reduced graphene oxide-based inverse spinel nickel ferrite nanocomposite. *Environmental Research*, 2022. **204**: p. 112029.
24. Narang, S.B., and K. Pubby, Nickel spinel ferrites: A review. *Journal of Magnetism and Magnetic Materials*, 2021. **519**: p. 167163.
25. Wang, L., et al., Adsorption capability for Congo red on nanocrystalline MFe_2O_4 (M = Mn, Fe, Co, Ni) spinel ferrites. *Chemical Engineering Journal*, 2012. **181**: p. 72–79.
26. Lingamdinne, L.P., et al., Biogenic reductive preparation of magnetic inverse spinel iron oxide nanoparticles for the adsorption removal of heavy metals. *Chemical Engineering Journal*, 2017. **307**: p. 74–84.
27. Malik, A.Q., et al., An overview on magnetic separable spinel as a promising materials for photocatalysis and waste water treatment. *ES Energy & Environment*, 2022. **19**: p. 744.
28. He, C., et al., Advanced organic recovery from municipal wastewater with an enhanced magnetic separation (EMS) system: Pilot-scale verification. *Water Research*, 2022. **217**: p. 118449.
29. Prot, T., et al., Magnetic separation and characterization of vivianite from digested sewage sludge. *Separation and Purification Technology*, 2019. **224**: p. 564–579.
30. Choi, Y.I., et al., Band gap-engineered ZnO and Ag/ZnO by ball-milling method and their photocatalytic and Fenton-like photocatalytic activities. *Applied Surface Science*, 2015. **356**: p. 615–625.
31. Hu, Y., et al., Defect-engineered WO3-x@ MoS2 hollow tube exhibiting enhanced Fenton-like and photocatalytic activities via electric field rearrangement and band alignment. *Applied Catalysis B: Environmental*, 2023. **320**: p. 122013.
32. SOUFI, A., et al., Spinel ferrites nanoparticles: Synthesis methods and application in heterogeneous Fenton oxidation of organic pollutants – a review. *Applied Surface Science Advances*, 2021. **6**: p. 100145.
33. Hauser, K., S.O. Hassou, and U. Riaz, Ultrasound-assisted photocatalytic degradation of Azo dyes under visible light irradiation using polythiophene-decorated CoFe2O4 nanohybrids. *Catalysts*, 2024. **14**(8): p. 495.
34. Del Río, L.A., and E. López-Huertas, ROS generation in peroxisomes and its role in cell signaling. *Plant and Cell Physiology*, 2016. **57**(7): p. 1364–1376.
35. Pecchillo Cimmino, T., et al., NOX dependent ROS generation and cell metabolism. *International Journal of Molecular Sciences*, 2023. **24**(3): p. 2086.
36. Zhang, Q., et al., Nano-ROS-generating approaches to cancer dynamic therapy: Lessons from nanoparticles. *Chemical Engineering Journal*, 2023. **457**: p. 141225.
37. Karbasi, M., et al., A review on Plasma electrolytic oxidation coatings for organic pollutant degradation: How to prepare them and what to expect of them? *Journal of Environmental Chemical Engineering*, 2023. **11**(3): p. 110027.

38. Wang, D., et al., Photoinduced acceleration of Fe^{3+}/Fe^{2+} cycle in heterogeneous FeNi-MOFs to boost peroxodisulfate activation for organic pollutant degradation. *Applied Catalysis B: Environmental*, 2023. **321**: p. 122054.
39. Tang, X., et al., Effects of different pretreatment strategies during porous carbonaceous materials fabrication on their peroxydisulfate activation for organic pollutant degradation: Focus on mechanism. *Chemical Engineering Journal*, 2023. **451**: p. 138576.
40. Wu, S., et al., Rapid activation of peroxymonosulfate with iron (III) complex for organic pollutants degradation via a non-radical pathway. *Water Research*, 2023. **233**: p. 119725.
41. Muhammad, Z., et al., Emerging cybersecurity and privacy threats to electric vehicles and their impact on human and environmental sustainability. *Energies*, 2023. **16**(3): p. 1113.
42. Chakraborty, R., et al., Nanocomposite-based smart fertilizers: A boon to agricultural and environmental sustainability. *Science of the Total Environment*, 2023. **863**: p. 160859.
43. Linghu, K., et al., Occurrence, distribution and ecological risk assessment of antibiotics in Nanming river: Contribution from wastewater treatment plant and implications of urban river syndrome. *Process Safety and Environmental Protection*, 2023. **169**: p. 428–436.
44. Grmasha, R.A., et al., Polycyclic aromatic hydrocarbons in the surface water and sediment along Euphrates River system: Occurrence, sources, ecological and health risk assessment. *Marine Pollution Bulletin*, 2023. **187**: p. 114568.
45. Sultan, M., et al., Organochlorine pesticides (OCPs) in freshwater resources of Pakistan: A review on occurrence, spatial distribution and associated human health and ecological risk assessment. *Ecotoxicology and Environmental Safety*, 2023. **249**: p. 114362.
46. Zhang, Y., et al., The distinct resistance mechanisms of cyanobacteria and green algae to sulfamethoxazole and its implications for environmental risk assessment. *Science of the Total Environment*, 2023. **854**: p. 158723.
47. Na, M., et al., Residues, potential source and ecological risk assessment of polycyclic aromatic hydrocarbons (PAHs) in surface water of the East Liao River, Jilin Province, China. *Science of the Total Environment*, 2023. **886**: p. 163977.
48. Ameen, F., and N. Majrashi, Recent trends in the use of cobalt ferrite nanoparticles as an antimicrobial agent for disability infections: A review. *Inorganic Chemistry Communications*, 2023. **156**: p. 111187.
49. Mondal, P., et al., Sustainable application of nanoparticles in wastewater treatment: Fate, current trend & paradigm shift. *Environmental Research*, 2023. **232**: p. 116071.
50. Katoch, G., et al., Sol-gel auto-combustion developed Nd and Dy co-doped Mg nano-ferrites for photocatalytic water treatment, electrocatalytic water splitting and biological applications. *Journal of Water Process Engineering*, 2023. **53**: p. 103726.
51. Kasavo, R.N., M. Bhaumik, and H.G. Brink, Removal of chromium from aqueous solution using a nanocomposite of nickel ferrite and polyaniline doped with 2-napthalene sulfonic acid. *Journal of Environmental Chemical Engineering*, 2023. **11**(3): p. 111229.

10 Role of Spinel and Inverse Spinel Ferrites in Plant Physiology and Growth Enhancement

Parvaiz Ahmad, Leonard Wijaya, Pravej Alam, Cengiz Kaya, and Mohammad Abass Ahanger

10.1 INTRODUCTION

The association of nanotechnology with agriculture has resulted in several developments concerning the growth and resistance of plants [1]. Various physiological parameters of plants were shown to be improved by ferrite magnetic nanomaterials, such as spinel and inverse spinel ferrites. Nanoferrite materials, for example, MFe_2O_4 where M accounts for Co, Zn, or Cu, etc., exhibit fine catalytic, magnetic, and adsorption properties, thereby becoming suitable candidates for agricultural applications [1, 2].

Spinel ferrites have also been reported to affect seed germination and root elongation along with nutrient uptake. These nanomaterials can also be considered as an eco-friendly source of chemical fertilizers. In this chapter, the author has described the effect of these nanomaterials on plant physiology and the future applications of these nanomaterials in precision agriculture. The crystal structure of spinel ferrites consists of a general formula AB_2O_4, where A and B are divalent and trivalent metal cations, respectively. In the structure of spinel ferrites, the B^{3+} ions occupy the positions of octahedral positions, whereas the A^{2+} ions occupy tetrahedral positions. However, for inverse spinel ferrites, the B^{3+} ions fill half of the tetrahedral sites. The other half of the B^{3+} ions together with A^{2+} ions fill the remaining half of the octahedral sites [3, 4]. These differences make crucial their respective chemical and biological interactions. Abdel hameed et al. studied the effect of zinc ferrite nanoparticles ($ZnFe_2O_4$ NPs) on the growth and physiological characteristics of *Pisum sativum* L. They further discussed how $ZnFe_2O_4$ NPs affected plant growth parameters, pigment content, mineral composition, and the mycorrhizal colonization of roots by arbuscular fungi. Their research proved that $ZnFe_2O_4$ NPs produced improved growth in plants due to the enhanced growth in both shoots and roots [5]. The most important finding was that chlorophyll and carotenoid content rose

 DOI: 10.1201/9781032665658-10

in the treated plants, thus proving $ZnFe_2O_4$ NPs have positive effects on photosynthesis processes. Greater content can also result in enhanced energy capture and subsequent growth. Mineral content in pea, for example, iron and zinc, was highly increased, showing an improvement in nutrient availability and uptake by the ferrite [6–7]. López-Moreno et al. evaluated cobalt ferrite ($CoFe_2O_4$) NPs to examine the impact they ascertain in *Lycopersicon lycopersicum*, the tomato plant, on its development and growth. It suggested the effect of NPs on seed germination, plant height, biomass production, and root length, among others, which account for the growth of the plants. The outcome observed that the actual effect was that $CoFe_2O_4$ NPs promoted growth in tomato plants under the controlled environment. The study was supplemented with increased elongation of roots and overall biomass production in treated tomato plants compared to the controls. Increased length of the root means that it occupies more water and nutrients, which keeps the plants generally healthier. The authors hypothesized that the magnetic properties of the $CoFe_2O_4$ NPs might have triggered the stimulation of root growth in these plants through action on cellular processes and signaling pathways related to nutrient uptake. Yue et al. investigated the molecular determinants underlying manganese ferrite ($MnFe_2O_4$) NPs induced early flowering in tomato plants [8]. This work gives an understanding of how $MnFe_2O_4$ NPs impact developmental processes in flowering plants. Flowering is a very critical phenological stage in the development of plants primarily because it directly influences fruit production and yield. The result showed that the application of $MnFe_2O_4$ NPs promoted the tomato plant flowering system to a very early time than normal by affecting the expression of marker genes for control of flowering. It seemed that the NPs somehow affected the hormonal regulation of the plants, especially at the gibberellins – a hormone that is involved in flower and bud formation. $MnFe_2O_4$ NPs activated the transcription factors responsible for the shift from vegetative to reproductive development, which accelerated flowering. Tombuloglu et al. tested the role of $MnFe_2O_4$ NPs on the growth of *Hordeum vulgare* L. [9]. This research explored the effects of NPs on plant growth, along with an added benefit of ascertaining the magnetic properties of plants post-treatment. Barley plants revealed improvements in growth and development due to $MnFe_2O_4$ NPs treatment shown by increased height, root length, and biomass. The scientists found that the treated plants also exhibited detectable magnetic properties, which indicates that the NPs had been absorbed and dispersed in the tissues of the barley plant. That which can, therefore, be made possible by exploiting these magnetic properties may include nutrient delivery with magnetic assistance or even plant health monitoring with the use of magnetic sensors. Despite all these proofs of the promotion of plant growth by ferrite NPs, there is also the toxicity issue. López-Luna et al. conducted a toxicological evaluation of wheat plants [10]. The purpose of this research was to analyze whether the application of nanomaterials could pose a risk of negative health effects for plants and their development. Results showed that even at low concentrations of $CoFe_2O_4$ NPs, it affected wheat growth insignificantly, but increasing concentration resulted in reduced plant biomass and chlorophyll, showing possible phytotoxicity at this point. Oxidative stress occurring in wheat plants due to NPs was also evidenced by high levels of reactive oxygen species (ROS) and lipid peroxidation. The study concluded that although $CoFe_2O_4$ NPs are effective at low concentrations, their application must be controlled to prevent any adverse effects on plant health. Owing to the aforementioned reports on the successful roles of ferrites for enhancing plant growth, the following section of this chapter will deal with what might be written concerning mechanisms involved: effects on seed germination, nutrient uptake and assimilation, stress tolerance, and defense mechanisms.

10.2 MECHANISMS OF ACTION IN PLANT PHYSIOLOGY

10.2.1 Seed Germination and Root Development

There have been reports of enhanced germination percentages of seeds by applying spinel ferrite coatings to the seeds and soil; it tends to increase the elongation of the roots generally. All these can be attributed to enhancing the bioavailability of essential micronutrients such as iron (Fe), magnesium (Mg), manganese (Mn), and zinc (Zn) for plants, which play critical roles in respiration, synthesis of chlorophyll, and the transfer of energy, among others [11]. Fe_3O_4, $MgFe_2O_4$, $MnFe_2O_4$, and $ZnFe_2O_4$ ferrites contain all the required micronutrients for plant growth and development. In particular, iron is the key component in cytochromes and other proteins containing heme groups that form the electron-transport chains essential for early plant development [12]. Zinc is also important for the production of auxins, which are plant hormones that control root growth and development [13]. Beyond nutrient availability, the special magnetic properties of spinel ferrites could influence other biological activities inside the seeds. The magnetic field developed by these nanomaterials could enhance water absorption by structuring water, which in turn would increase water retention within the seed cell [14]. This would enhance metabolic activity, thereby accelerating germination. Spinel ferrites could also alter the physicochemical properties of the soil to allow uptake of other important elements such as phosphorus for further support of early plant development. Additionally, increased root length and density with the application of spinel ferrite may lead to enhanced uptake of nutrients and water, thereby improving plant survival under suboptimal conditions [15]. These improved root architectures further add to the vigor, stress tolerance, and productivity of the plant (Figure 10.1).

Perea-Vélez et al. reported that the $CoFe_2O_4$ NPs affected the growth of wheat plants and enhanced toxicity at higher concentrations. Inhibitory effects were manifested with the germination and root and shoot lengths reduced. The essence of the study lies in the identification that it manifests a dose-dependent nature of the effect involved. The stimulation can be observed in some of the growth parameters at low concentrations [16]. Ilyas et al. examined carbon nanodots and $CoFe_2O_4$ NPs to promote growth and antioxidant activity in drought-stressed maize inbred lines, indicating its future possibilities in enhancing plant stress tolerance and resilience [17]. The potential of $CuFe_2O_4$ NPs was examined as an added supplement during the hydroponic cultivation of cucumber plants, and investigations were made based on the effects of $CuFe_2O_4$ NPs on plant growth, nutrient uptake, and overall physiological performance [18]. A potential boost in photosynthesis, total yield increase, and plant

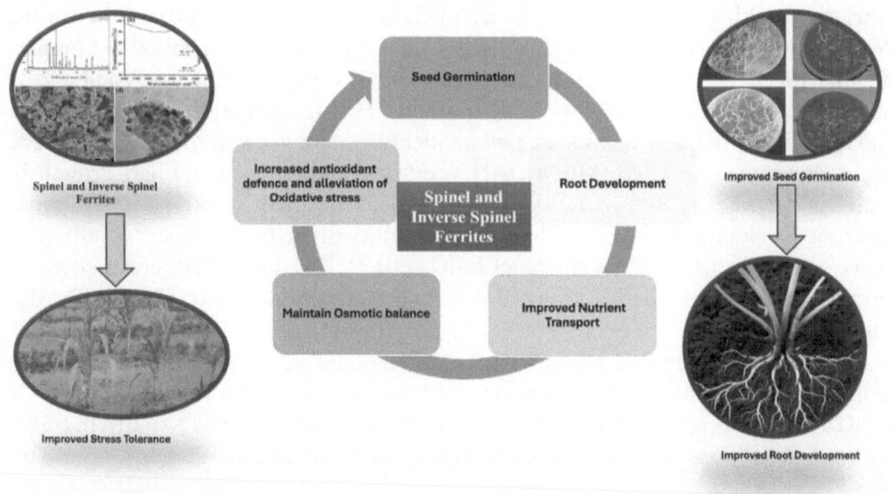

FIGURE 10.1 Demonstrating the importance of spinel and inverse spinel ferrites in plant developmental processes.

elongation that enhanced the agricultural productivity of wheat (*Triticum aestivum* L.) was observed when treated with the Ce-Mn ferrite nanocomposite [19]. $ZnFe_2O_4$ NPs synthesized by the green approach favor seed germination and development of parsley. The green synthesis method minimizes the toxicity level due to an increased growth of the root and shoot. Proper doses of NPs have been found to enhance nutrient assimilation in plants, which further helps in increasing growth [20].

10.2.2 Nutrient Uptake and Assimilation

The most vital role of spinel ferrites in agriculture is that they improve nutrient uptake. Nutrient uptake, especially by essential micronutrients, is through ferrites, particularly iron, zinc, and copper. Iron deficiency generally leads to limited growth and productivity among crops. There are chlorosis conditions in which plants cannot produce enough chlorophyll. Ferrite NPs serve to alleviate this deficiency by being used as slow-releasing sources of iron [21]. These iron NPs have been found to release gradually into the plants for easy assimilation, hence providing the continuation of ferric ion supply over time.

However, nanoferrites enhance the surface area for nutrient adsorption to take place; thus, increasing their efficiency in making nutrients in soil. Ferrites can serve as carriers for other essential nutrients. This improves the nutrient profile available to plants [22]. For example, ferrite NPs can be functionalized or co-loaded with micronutrients such as zinc or magnesium to transport the latter into the root zone and then further into the cells of the plant. In addition to the direct addition of nutrients, ferrites can also change the soil environment in such a way that mobilizes existing nutrients, making them more available. Their catalytic properties can enhance the decomposition of organic matter in soils, releasing bound nutrients in complex organic forms. They may also induce root exudates that contribute to solubilizing phosphorus and other poorly available nutrients. Ferrites can boost both the quality and quantity of nutrient uptake, enhancing healthy plant growth, tolerance to stress, and increasing crop yield. $MnFe_2O_4$ NPs enhance nutrient uptake and growth in maize under drought stress conditions by increasing antioxidant activity. They stimulate root development, thereby enhancing the ability of the plants to access nutrients, hence elevating the health of the plant in its entirety [23]. Fe_3O_4 NPs enhance nutrient uptake by barley as they promote the uptake of vital nutrients. These NPs also control the expression of photosynthetic genes, thus making the crop more effective in its utilization of the obtained nutrients [24]. The NPs promote early flowering in tomatoes, concurrently increasing nutrient acquisition and assimilation that supports growth [25]. $MnFe_2O_4$ NPs increase the expression of genes linked to nutrient transport and metabolism in tomatoes, ensuring enhanced uptake of nutrients.

10.2.3 Stress Tolerance and Defense Mechanisms

Spinel ferrites increase plant resistance to abiotic stresses, such as drought, salinity, and heavy metal toxicity. They strengthen the antioxidant defense system of the plant and encourage the activity of many major antioxidant enzymes such as superoxide dismutase (SOD) and catalase (CAT), capable of neutralizing dangerous ROS that are created under the influence of stress [26, 27] (Figure 10.2).

Thus, they reduce oxidative damage to cellular structures and make them resistant to environmental stresses. Nanoferrites promote the synthesis of osmoprotectants, which are the molecules that help plants retain water and keep their cells stable

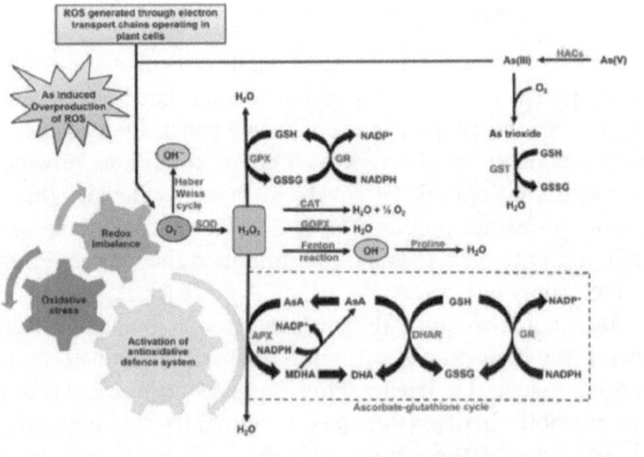

FIGURE 10.2 Schematic outline of antioxidant defense machinery in plants that counter the negative effects of redox imbalance and oxidative stress caused by reactive oxygen species (ROS) production [28].

during drought or salinity stress. They also minimize heavy metal toxicity because they prevent noxious metals locked in the soil from being absorbed by plant roots. Furthermore, the possible influence of spinel ferrites on stress-related hormones such as abscisic acid (ABA) can enhance the defense mechanisms and stress adaptability of plants.

The gamma-irradiated synthesized novel Cs/Ag/$MnMgFe_2O_4$ nanocomposite limits stress in plants under cadmium (Cd) stress by reducing the accumulation of cadmium, improving photosynthetic parameters, and enhancing antioxidant and non-antioxidant enzymes, making it to be considered as a potential stress management tool in plants [29].

The application of gallic acid along with $ZnFe_2O_4$ NPs in wheat improves root structure and enhances antioxidant defense to overcome the adverse effects of salinity stress. This system promotes the synthesis of stress-related proteins and thus enhances the tolerance level of the plant toward salt stress [30].

10.3 ENVIRONMENTAL AND AGRICULTURAL APPLICATIONS

10.3.1 Role in Sustainable Agriculture

Ferrites are part of the spinel and inverse spinel families. These nanomaterials have tremendous potential in sustainable agriculture because of their magnetic, electrical, and catalytic properties. These materials drastically improve nutrient delivery as slow-release carriers for essential nutrients such as iron, zinc, and magnesium, which are specifically deficient in soils. Ferrites reduce the demand for often recurring applications of chemical fertilizers due to the gradual release of these nutrients in bioavailable forms; hence, they make soil degradation and pollution more

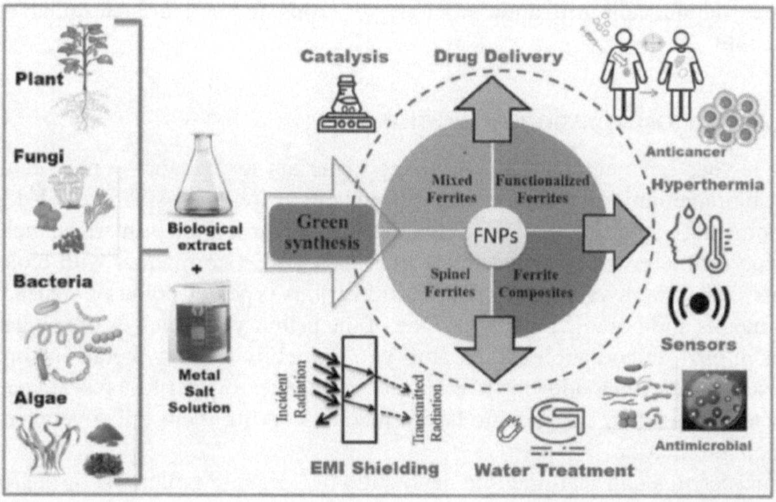

FIGURE 10.3 Synthesis and various applications of spinel and inverse spinel ferrites [31].

controllable. They also promote better plant development with improved seed germination and root development.

Spinel ferrites synthesized from a low-cost, biodegradable, and green route are environmentally benign as shown in Figure 10.3. Due to nanotechnology, their application improves the use efficiency of agricultural inputs while minimizing the impacts of farming practices on the environment, such as runoff into water bodies and deficiency of nutrients in plants. The use of spinel and inverse spinel ferrites is expected to promote better nutrient utilization, tolerance of stresses in plants, and aid in the reduction of the ecological impacts associated with agriculture. These materials might improve crop yields while maintaining ecological balance and long-term soil health.

$MnFe_2O_4$ NPs synthesized via a green approach have proven to be nanofertilizers for *Cucurbita pepo*, as their application has increased plant growth due to better nutrient assimilation and soil health. Their eco-friendly synthesis makes them ideal for sustainable agriculture practices [32].

The principal application of $MnFe_2O_4$ NPs is through micronutrient fertilizers, promoting the concept of green agriculture. These NPs enhance growth and yield, thereby facilitating the production of greener crops [33].

$MgFe_2O_4$ NPs used as priming improved the elemental composition and growth of *Capsicum chinense*, presenting a potential application of nanofertilizers in sustainable agriculture. The NPs enhance nutrient assimilation and support healthy plant development [34]. Citrate-coated $CoFe_2O_4$ NPs enhance the biofortification of wheat through increased availability and uptake of nutrients. Nano-enabled fertilization supports practices that promote sustainable agriculture by improving wheat growth and grain quality [35]. The application of arbuscular mycorrhizal fungi and nano-$ZnFe_2O_4$ NPs enhances the growth and nutritional content of pea plants with the increased primary metabolites and activity of enzymes. In this way, this method enhances the nutrient utilization efficiency of crops and supports sustainable agriculture [36].

10.4.2 Phytoremediation Applications

Spinel ferrites have emerged as promising materials for enhancing phytoremediation, a bioremediation process that utilizes plants to remove, stabilize, or degrade environmental contaminants from soil and water. Their unique properties, including magnetic characteristics, large surface area, and ability to interact with biological systems, make them valuable in addressing various types of pollution, particularly heavy metals and organic pollutants. One of the primary advantages of using spinel ferrites in phytoremediation is their ability to immobilize heavy metals in contaminated soils [37]. By modifying the surface properties of ferrites, researchers can enhance their affinity for specific heavy metals, making them effective agents for soil decontamination.

Spinel ferrites can facilitate the degradation of organic pollutants through catalytic processes. Some ferrite NPs possess inherent catalytic activity, enabling them to break down complex organic compounds, such as pesticides or hydrocarbons, into less harmful substances [38]. When applied in conjunction with phytoremediation, these ferrites can help plants metabolize or detoxify organic contaminants, leading to a more comprehensive remediation approach. Magnetic NPs, including ferrites, enhance phytoremediation by improving the uptake and transport of contaminants in plants. Their magnetic properties allow for the easy recovery of contaminants from the environment, making them an efficient tool for environmental cleanup [39]. Nanoferrites play a significant role in the removal of environmental contaminants by acting as adsorbents and catalysts. Their ability to degrade toxic substances through photocatalysis and adsorption supports their use in sustainable phytoremediation practices [40]. $ZnFe_2O_4$ NPs synthesized via green methods show high efficiency in the photocatalytic degradation of methylene blue, a common environmental pollutant. Their use in phytoremediation highlights the potential of nanoferrites in removing dyes and other toxic compounds from water and soil [41]. $Ni_xZn_{(1-x)}Fe_2O_4$ NPs show highly efficient photocatalytic degradation of diclofenac, a contaminant to aquatic systems. The eco-toxicity of the NPs was found to be very minimal, thus safe for phytoremediation processes in environmental remediation [42]. The magnetic ferrites supported on porous carbons, $NiFe_2O_4$, $CoFe_2O_4$, $ZnFe_2O_4$, and Fe_3O_4, are excellent for the adsorption of toxic dyes released in the environment. These nanomaterials are part of phytoremediation systems, improving the removal of contaminants without compromising the safety of the environment [43].

10.4 CONCLUSION

An overview of ferrite magnetic nanomaterials' role in the development of agriculture is well furnished in this introduction. The spinel ferrites of MFe_2O_4, where M could be Co, Zn, or Cu, possess catalytic, magnetic, and adsorption properties. These nanomaterials have primarily shown functions of enhancing seed germination, root elongation, and nutrient uptake and have provided the world with a scope of saving chemical fertilizers, a green alternative. Some of the key studies indicate that several ferrite nanoparticles have a positive impact on plant physiology. The chapter elucidates mechanisms whereby ferrite nanomaterials can facilitate growth in plants – a mechanism or process that involves improved nutrient bioavailability and enhanced absorption of water due to the magnetic properties. These ferrite nanomaterials also ensure stress tolerance by increasing antioxidant defense and reducing heavy metal toxicity. By considering the role of ferrites in sustainable agriculture, their role as slow carriers of essential nutrients to provide better nutrition is envisioned in reducing dependency on chemical fertilizers and promoting eco-friendly agricultural practices. Ferrite nanomaterials are shown to be promising candidates for sustainable and efficient crop production, besides supporting better assimilation of nutrients and soil health

REFERENCES

1. Spanos, A., Athanasiou, K., Ioannou, A., Fotopoulos, V., & Krasia-Christoforou, T. (2021). Functionalized magnetic nanomaterials in agricultural applications. *Nanomaterials*, *11*(11), 3106.
2. Thakur, A., & Thakur, P. (2023). Ferrite nanoparticles for agriculture-related activity. In *Applications of Nanostructured Ferrites* (pp. 315–330). Woodhead Publishing.
3. Tatarchuk, T., Bououdina, M., Judith Vijaya, J., & John Kennedy, L. (2017). Spinel ferrite nanoparticles: synthesis, crystal structure, properties, and perspective applications. In *Nanophysics, Nanomaterials, Interface Studies, and Applications: Selected Proceedings of the 4th International Conference Nanotechnology and Nanomaterials (NANO2016), August 24–27, 2016, Lviv, Ukraine* (pp. 305–325). Springer International Publishing.
4. Rafiq, M. A., Javed, A., Rasul, M. N., Nadeem, M., Iqbal, F., & Hussain, A. (2021). Structural, electronic, magnetic and optical properties of AB_2O_4 (A = Ge, Co and B = Ga, Co) spinel oxides. *Materials Chemistry and Physics*, *257*, 123794.
5. Abdelhameed, R. E., Abu-Elsaad, N. I., Abdel Latef, A. A. H., & Metwally, R. A. (2021). Tracking of zinc ferrite nanoparticle effects on pea (*Pisum sativum* L.) plant growth, pigments, mineral content and arbuscular mycorrhizal colonization. *Plants*, *10*(3), 583.
6. Tombuloglu, H., Alsaeed, M., Slimani, Y., Demir-Korkmaz, A., Tombuloglu, G., Sozeri, H., ... Ercan, I. (2023). Formulation of manganese zinc spinel ferrite ($Mn_{0.5}Zn_{0.5}Fe_2O_4$) nanoparticles for the growth promotion of plants. *Journal of Soil Science and Plant Nutrition*, *23*(3), 3561–3574.
7. López-Moreno, M. L., Avilés, L. L., Pérez, N. G., Irizarry, B. Á., Perales, O., Cedeno-Mattei, Y., & Román, F. (2016). Effect of cobalt ferrite ($CoFe_2O_4$) nanoparticles on the growth and development of *Lycopersicon lycopersicum* (tomato plants). *Science of the Total Environment*, *550*, 45–52.
8. Yue, L., Feng, Y., Ma, C., Wang, C., Chen, F., Cao, X., ... Xing, B. (2022). Molecular mechanisms of early flowering in tomatoes induced by manganese ferrite ($MnFe_2O_4$) nanomaterials. *ACS Nano*, *16*(4), 5636–5646.
9. Tombuloglu, H., Tombuloglu, G., Slimani, Y., Ercan, I., Sozeri, H., & Baykal, A. (2018). Impact of manganese ferrite ($MnFe_2O_4$) nanoparticles on growth and magnetic character of barley (*Hordeum vulgare* L.). *Environmental Pollution*, *243*, 872–881.
10. López-Luna, J., Camacho-Martínez, M. M., Solís-Domínguez, F. A., González-Chávez, M. C., Carrillo-González, R., Martinez-Vargas, S., ... Cuevas-Díaz, M. C. (2018). Toxicity assessment of cobalt ferrite nanoparticles on wheat plants. *Journal of Toxicology and Environmental Health, Part A*, *81*(14), 604–619.
11. Mahesh, B. (2023). A comprehensive review on current trends in greener and sustainable synthesis of ferrite nanoparticles and their promising applications. *Results in Engineering*, *21*, 101702.
12. Kranz, R. G., Richard-Fogal, C., Taylor, J. S., & Frawley, E. R. (2009). Cytochrome c biogenesis: mechanisms for covalent modifications and trafficking of heme and for heme-iron redox control. *Microbiology and Molecular Biology Reviews*, *73*(3), 510–528.
13. Karimi, N., Goltapeh, E. M., Amini, J., Mehnaz, S., & Zarea, M. J. (2021). Effect of *Azospirillum zeae* and seed priming with zinc, manganese and auxin on growth and yield parameters of wheat, under dryland farming. *Agricultural Research*, *10*, 44–55.
14. Kornarzyński, K., Sujak, A., Czernel, G., & Wiącek, D. (2020). Effect of Fe_3O_4 nanoparticles on germination of seeds and concentration of elements in *Helianthus annuus* L. under constant magnetic field. *Scientific Reports*, *10*(1), 8068.

15. Răcuciu, M., Creangă, D. E., & Oprică, L. (2023). Nanosized ferrites in environmental sciences. In *Applications of Nanostructured Ferrites* (pp. 331–351). Woodhead Publishing.
16. Perea-Vélez, Y. S., Carrillo-González, R., González-Chávez, M. D. C. A., Vangronsveld, J., Monasterio, I. O., & Maruri, D. T. (2023). Citrate-coated cobalt ferrite nanoparticles for the nano-enabled biofortification of wheat. *Food & Function, 14*(9), 4017–4035.
17. Ilyas, M. Z., Park, H., Baek, Y. S., Sa, K. J., Kim, M. J., & Lee, J. K. (2023). Efficacy of carbon nanodots and manganese ferrite ($MnFe_2O_4$) nanoparticles in stimulating growth and antioxidant activity in drought-stressed maize inbred lines. *Plants, 12*(16), 2922.
18. Abu-Elsaad, N. I., & Abdel hameed, R. E. (2019). Copper ferrite nanoparticles as nutritive supplement for cucumber plants grown under hydroponic system. *Journal of Plant Nutrition, 42*(14), 1645–1659.
19. Zarinkoob, A., Esmaeilzadeh Bahabadi, S., Rahdar, A., Hasanein, P., & Sharifan, H. (2021). Ce-Mn ferrite nanocomposite promoted the photosynthesis, fortification of total yield, and elongation of wheat (*Triticum aestivum* L.). *Environmental Monitoring and Assessment, 193*, 1–12.
20. Korotkova, A. M., Polivanova, O. B., Gavrish, I. A., Baranova, E. N., & Lebedev, S. V. (2019, October). Green synthesis of zinc based nanoparticles zinc ferrite by *Petroselinum crispum*. In *IOP Conference Series: Earth and Environmental Science* (Vol. 341, No. 1, p. 012175). IOP Publishing.
21. Haydar, M. S., Ghosh, D., & Roy, S. (2024). Slow and controlled release nanofertilizers as an efficient tool for sustainable agriculture: Recent understanding and concerns. *Plant Nano Biology, 7*, 100058.
22. Hieu, D. M., Hoa, N. T., Oanh, H. T. M., & Duy, T. N. (2021). Ferrite (Fe_3O_4) nanoparticle in soil stimulates the plant growth in peas and Bok Choy. *Journal of Science and Technology, 31*(4), 38–45.
23. Ilyas, M. Z., Park, H., Baek, Y. S., Sa, K. J., Kim, M. J., & Lee, J. K. (2023). Efficacy of carbon nanodots and manganese ferrite ($MnFe_2O_4$) nanoparticles in stimulating growth and antioxidant activity in drought-stressed maize inbred lines. *Plants, 12*(16), 2922.
24. Tombuloglu, H., Slimani, Y., Tombuloglu, G., Almessiere, M., & Baykal, A. (2019). Uptake and translocation of magnetite (Fe_3O_4) nanoparticles and its impact on photosynthetic genes in barley (*Hordeum vulgare* L.). *Chemosphere, 226*, 110–122.
25. Yue, L., Feng, Y., Ma, C., Wang, C., Chen, F., Cao, X., ... Xing, B. (2022). Molecular mechanisms of early flowering in tomatoes induced by manganese ferrite ($MnFe_2O_4$) nanomaterials. *ACS Nano, 16*(4), 5636–5646.
26. Zia-ur-Rehman, M., Anayatullah, S., Irfan, E., Hussain, S. M., Rizwan, M., Sohail, M. I., ... Alharby, H. F. (2023). Nanoparticles assisted regulation of oxidative stress and antioxidant enzyme system in plants under salt stress: A review. *Chemosphere, 314*, 137649.
27. Movahedi, A., Taghimolaee, M., Ghayebpanah, S., Monajemi, F., Ghanbari, M., & Farhazadi, H. N. (2024). A review of the effect of nanoparticles on reactive oxygen species production in different plants. *Journal of Engineering and Applied Research, 1*(1), 131–163.
28. Mondal, S., Pramanik, K., Ghosh, S. K., Pal, P., Ghosh, P. K., Ghosh, A., & Maiti, T. K. (2022). Molecular insight into arsenic uptake, transport, phytotoxicity, and defense responses in plants: a critical review. *Planta, 255*(4), 87.
29. Maksoud, M. A., Bekhit, M., El-Sherif, D. M., Sofy, A. R., & Sofy, M. R. (2022). Gamma radiation-induced synthesis of a novel chitosan/silver/Mn-Mg ferrite nanocomposite and its impact on cadmium accumulation and translocation in brassica plant growth. *International Journal of Biological Macromolecules, 194*, 306–316.

30. Shao, Q., Ren, L., Ramzan, M., Hussain, M. B., Datta, R., Almoallim, H. S., ... Ehsan, A. (2024). Combined effect of gallic acid and zinc ferrite nanoparticles on wheat growth and yield under salinity stress. *Scientific Reports, 14*(1), 12854.
31. Mahesh, B. (2023). A comprehensive review on current trends in greener and sustainable synthesis of ferrite nanoparticles and their promising applications. *Results in Engineering, 21*, 101702.
32. Shebl, A., Hassan, A., Salama, D., Abd El-Aziz, M. E., & Abd Elwahed, M. (2019). Green synthesis of manganese zinc ferrite nanoparticles and their application as nanofertilizers for *Cucurbita pepo* L. *Beilstein Archives, 2019*(1), 45.
33. Huang, X., Wang, X., Liu, X., Cheng, L., Pan, J., & Yang, X. (2024). Nanotechnology in agriculture: Manganese ferrite nanoparticles as a micronutrient fertilizer for wheat. *Plants, 13*(10), 1395.
34. Coria-Téllez, A. V., Guerrero, C. A., Castro, C. A., Rodríguez Mendiola, M. A., & Tiwari, D. K. (2023). Elemental microanalysis of *Capsicum chinense* plants treated with magnesium doped iron ferrite nanopriming. *Microscopy and Microanalysis, 29*(Supplement_1), 28–29
35. Perea-Vélez, Y. S., Carrillo-González, R., González-Chávez, M. D. C. A., Vangronsveld, J., Monasterio, I. O., & Maruri, D. T. (2023). Citrate-coated cobalt ferrite nanoparticles for the nano-enabled biofortification of wheat. *Food & Function, 14*(9), 4017–4035.
36. Metwally, R. A., & Abdelhameed, R. E. (2024). Co-application of arbuscular mycorrhizal fungi and nano-ZnFe2O4 improves primary metabolites, enzymes and NPK status of pea (*Pisum sativum* L.) plants. *Journal of Plant Nutrition, 47*(3), 468–486.
37. Tian, Y., Qu, G., Xu, R., Liu, X., & Jin, C. (2024). Iron-based materials for immobilization of heavy metals in contaminated soils: a critical review. *Journal of Environmental Chemical Engineering, 12*(5), 113741.
38. Cheng, Y., Zhang, S., Wang, Z., Wang, B., You, J., Guo, R., & Zhang, H. (2023). Review on spinel ferrites-based materials (MFe_2O_4) as photo-Fenton catalysts for degradation of organic pollutants. *Separation and Purification Technology, 318*, 123971.
39. Nouri, M. (2024). Magnetic nanotechnology-based sensors for environmental contaminants' detection. In *Nanotechnology-Based Sensors for Detection of Environmental Pollution* (pp. 155–174). Elsevier.
40. Kaymak, G. (2023). Role of nanoferrites for the removal of contaminants from environment. In *Biogenic Nanomaterial for Health and Environment* (pp. 202–209). CRC Press.
41. Din, M. I., Jabbar, S., Najeeb, J., Khalid, R., Ghaffar, T., Arshad, M., ... Ali, S. (2020). Green synthesis of zinc ferrite nanoparticles for photocatalysis of methylene blue. *International Journal of Phytoremediation, 22*(13), 1440–1447.
42. Mohan, H., Sathya, P. M., Vadivel, S., Ha, G. H., Oh, H. S., Kim, G., ... Shin, T. (2022). Highly efficient visible light photocatalysis of $Ni_xZn_{1-x}Fe_2O_4$ ($x = 0, 0.3, 0.7$) nanoparticles: Diclofenac degradation mechanism and eco-toxicity. *Chemosphere, 301*, 134699.
43. Nguyen, D. T. C., Jalil, A. A., Hassan, N. S., Nguyen, L. M., Nguyen, D. H., & Van Tran, T. (2024). Synthesis of magnetic MFe_2O_4 (M = Ni, Co, Zn, Fe) supported on porous carbons derived from *Bidens pilosa* weed and their adsorptive comparison of toxic dyes. *Chemosphere, 358*, 142087.

11 Gas-Sensing Applications of Ferrites

Manjeet Kumar, Aditya Nath Mishra, and Chandra Kumar

11.1 INTRODUCTION

The increasing demand for gas sensor technologies has been employed in various areas, including environmental protection, industrial explosive gas detection for safety purposes, fermentation control in food production, medical diagnostics, and household combustible gas detection for safety purposes [1–3]. The global gas sensor market was estimated to be around 1.7 billion USD in 2012, with growth at a compound annual rate of 5.1%, continuously emphasizing the necessity for advanced and efficient materials for gas detection [4]. Spinel ferrites among the different materials investigated for gas sensing have emerged as promising alternatives with versatile structural properties and the ability to be tailored for targeted sensing applications. Ferrites, with the general formula AB_2O_4, are complex oxides that indeed hold considerable potential in gas-sensing applications because of their unique crystal structures and cation distribution between tetrahedral and octahedral sites [5]. Generally, these materials were traditionally recognized for exhibiting ferromagnetic properties in high-frequency applications but have more recently been examined in gas sensing. The interaction of ferrite materials with gases causes electrical resistance variations, which is a very simple and efficient method for the detection of concentration [6–8]. Ferrites have great potential to detect different gases such as volatile organic compounds (VOCs), H_2, NH_3, CO, and even SO_x, while showing selective sensitivity to sulfur oxides with $MgFe_2O_4$ [9–12]. The versatility of the composition of spinel ferrites as hybrid systems enables tuning their gas-sensing properties, including sensitivity, response time, and selectivity, through the incorporation of different transition and post-transition metal cations [13]. This is critical in terms of increasing demand for sensors with lower concentrations of gases that respond faster and have improved stability over long periods of time, especially for remote sensing in smart devices [3, 8]. Moreover, the stability and reversibility of these materials under various environmental conditions have made them promising candidates for real-world applications in various sectors. This chapter aims to discuss ferrites as gas sensors by emphasizing their crystal structure, synthesis techniques, and mechanisms of gas sensing. It also covers the present progress in research studies on spinel ferrites regarding the detection of gases such as H_2S, VOCs, CO, NH_3, and H_2S, among others. We now

DOI: 10.1201/9781032665658-11

discuss the possibility that ferrites might help to improve technologies of gas sensors through optimization of their physical and chemical properties, playing an essential part in advancing the gas sensor market contributing to safer, more efficient detection systems.

Ferrites have grown to become an important component in the development of gas sensors due to their unique combination of electrical, magnetic, and catalytic properties. Though normally semiconducting oxides, spinel ferrites are highly sensitive and selective for gas detection, including CO_2, NO_2, NH_3, and H_2S even at very low concentrations. Doping ferrites with specific metal ions enhance their sensitivity and selectivity. This allows for more accurate detection of gases. Such adaptability makes ferrite-based sensors suitable for use in environmental monitoring, industrial safety, and healthcare sectors. The major advantage of ferrites is their chemical stability, by which they are suitable for operation under extreme conditions, such as high temperatures and aggressive atmospheres. This is an important consideration for applications in automotive exhaust monitoring and other chemical processing and industrial plants. Ferrites are manufactured at low cost, thus making them promising for mass production of gas sensors that ensure durability and require little maintenance. Ferrites also hold promise for hybrid gas sensors, where besides detection capability, the functionality also allows for magnetic manipulation, which would enhance the sensor response and recovery rates. Ferrite-based sensors are easy to apply in advanced technologies such as wireless sensor networks and wearable devices for real-time gathering of data and remote monitoring. This makes them an important auxiliary tool for providing environmental safety, public health, and sustainable development by enhancing the monitoring of air quality and detecting hazardous gases in many areas.

11.2 VARIOUS CHARACTERISTIC PARAMETERS OF GAS-SENSING MECHANISM

11.2.1 Structure of Pore

The insulating grains of the ferrite material exhibit voids known as pores, and these pores can be characterized into three subgroups [14, 15]:

i. The pores that begin from the surrounding surface of the material then pass through a certain volume and end at the same initial surface of the material.
ii. These pores enter deep into the bulk of the material body and reach the other end of the material.
iii. The pores that are created within the bulk body but don't have an end at the other side of the material.

Based on the geometry of the pores such as split, cylindrical, cubical, rectangular, and various cross-sections, the size of the pore can be determined. There are mainly two types of pores: (a) micropores and mesopores. The porosity (P) is one of the important parameters in the gas-sensing mechanism, and it can be defined as:

$$P = \frac{V_P}{V} \tag{11.1}$$

Here, V is the volume of the bulk material, and V_P is the volume of the pores inside the material. Another important characteristic in gas sensing is a specific surface (s). It is defined as the surface of the bulk material to the total mass of the body and it can be represented as:

$$s = \frac{S}{m} \tag{11.2}$$

The main dependence of the gas sensor is on the specific surface. Within definite limits, a large surface area provides high sensitivity. It does not make it an easy job to manage grain size, porosity, and specific pore surface area in the production of sensors. The basic characteristic of the porous material is pore allocation by size. This allocation controls sensor sensitivity [16].

11.2.2 Relationship between Sensor Resistance and Gas Concentration

This property describes the dependence of sensor resistance, or conductance, on gas concentration (C_g). For a dilute concentration of the gas in an atmosphere of air at a fixed constant temperature, conductance (G) is given by [16]:

$$G = KC_g^{\pm} \tag{11.3}$$

where K and α are constants and stand for gas concentration parameters in air.

11.2.3 Gas Sensor Sensitivity and Its Determinants

The sensitivity (S_g) of the sensor is taken as the degree of influence of a gas under consideration on the resistance of the sensor. It is usually defined by a ratio of the resistance in the air, R_a, to the resistance, R_g, given in a gas with concentration (C_g)) [17] :

$$S_g(\%) = \frac{R_a}{R_g} \times 100 \tag{11.4}$$

The sensitivity depends on the concentration of gases and properties of the sensing material [18, 19]. With the presence of an oxidizing gas, there is an increase in resistance in the sensor, whereas it drops with the presence of a reducing gas. This means that the sensor's response is both from the oxidizing and reducing gases [20]. Nevertheless, this definition has several weaknesses [21, 22], which include:

(a) We assume that air also possesses some resistance as a basis. Hence, it is difficult to make comparisons between the parameters of the sensor while resistance varies with different ranges.
(b) Sensitivity, as defined in this context, is a specific characteristic of an individual sensor at a given concentration of the gas; thus, it is contingent upon concentration and does not represent a universal parameter.

The sensitivity index provided in Equation (11.3) is a more general sensor parameter insensitive to the measuring process. In or near inter-grain contacts, the combustible gas is adsorbed by negatively charged oxygen adsorbates as seen in Equation (11.4), and so $\frac{R_a}{R_g}$ gives the value. Sensor sensitivity hence depends on the steady-state coverages of oxygen adsorbates as compared to that in the air. Moreover, sensitivity can be defined as the ratio of the variation of resistance, ” R, in the test gas to the resistance in air, R_a [23]:

$$S(\%) = \frac{''R}{R_a} \times 100 \tag{11.5}$$

The gas sensors are sensitive to pore size, porosity, and specific surface area. Although it is challenging, the manufacturing process, appropriate temperature ranges and pore-forming additives can overcome this problem. Moreover, the higher specific surface area increases the sensitivity, but that is up to certain limits [24–26].

11.2.4 Dynamic Response Behavior of Gas Sensors

This implies the general response curve behavior in relation to resistance and time as a function of various concentrations of gas. The test for the response is done by taking the test circuit from clean air to a given concentration of gas, under controlled temperatures and gas concentrations [27]. These properties enable the understanding of the rise time and fall time associated with the sensor. The rise time is the time taken for the sensor to detect and respond to the availability of the gas. In contrast, the falling time specifies the time taken by the sensor to get back to its baseline after the removal of the test gas [28].

11.2.5 Hysteresis Effects in Gas Sensor Performance

Hysteresis is described as the difference in the output of a sensor for a certain input, which is applied in both forward and backward directions. The reasons for hysteresis are varied and depend on the type of sensor. In magnetic sensors, it arises due to the delay in the orientation of dipoles, while in semiconductor sensors, it arises from the slow traps present in injection-type systems [29]. Hysteresis is defined as the change in steady sensitivity values with increasing and decreasing gas concentrations. Based on the principle of the ideal sensor, the ideal sensor should actually show less hysteresis [30].

11.2.6 Gas Sensor Selectivity and Optimization

Selectivity refers to the ability to specifically recognize a particular gas among other gases, and it closely correlates with the operating temperature [31]. A response toward a mixture of gases can be viewed as an effect or average of all the gases that exist; therefore, it is essential that the response remains selective, showing a primary response to one specific gas. In the context of gas adsorption on the sensor grains' surface, always accompanied by a chemical reaction, gas sensitivity is described. The activation energy of the chemical reaction depends linearly on temperature; therefore, sensitivity is influenced by the operating temperature of the sensor. Selectivity, operating temperature, and sensitivity often cannot exist without each other. The following methods may increase selectivity and sensitivity [32]:

i. Temperature adjustment
ii. Catalysts and dopants
iii. Addition of specific additives on a larger grain surface
iv. Use of filters

Moreover, sensitivity and selectivity can be enhanced using electrode configuration modification and thickness of the sensing layer. The geometry and structure of the electrodes, along with locations and distances relative to the sensing layer, determine the overall sensitivity and specificity of the sensor [33].

11.2.7 Impact of Operating Temperature on Gas Sensor Performance

Obviously, this is the operating temperature – a determining parameter for affecting the sensitivity of a ceramic gas sensor. The changes in the resistor resistance to a given gas and the sensor's response depend on the activation processes involved, meaning the speed of the chemical reaction occurring on the grain surface and the diffusion speed of detected molecules toward the surface. The activation energy for the chemical reaction is relatively high, so at lower temperatures, the response is limited by a slower reaction rate, while at higher temperatures, it is limited by the slower diffusion of the gas molecules. In that situation, the sensor's response is only optimal at intermediate temperatures, in which both factors have speeds that are about equal. Each gas has a particular temperature at which it has the most sensitive threshold of response, depending on the kind of gas, chemical composition, additives, and catalysts [34].

Generally, with most sensors, sensitivity increases with temperature, peaks, and then decreases as temperature is increased. The sensing mechanism in metal–oxide–semiconductor sensors is due to the interaction of target gases with the sensor surface. Typically, at low temperatures, gas molecules have not had enough thermal energy to interact with the oxygen species adsorbed on the surface. Thus, this results in a lower reaction rate and therefore reduced sensitivity. Two primary reasons make the sensitivity grow with an increase in temperature:

i. Increasing thermal energy increases the reaction rate by overcoming of activation energy.

At high temperatures, the abstraction of the oxygen species adsorbed from O_2 to O_2^- leads to an increase in electron concentration and promotes a reaction between the target gas and O_2^-.

There, sensitivity decreases again at the peak value since it is very difficult to adsorb exothermic gases at higher temperatures. Thus, there should be an optimal temperature at which sensitivity increases [35, 36].

11.3 GAS-SENSING MECHANISM IN FERRITES

Ferrite-based gas sensors operate on a well-established mechanism of semiconductor gas sensors whereby, after chemisorption of oxygen species on the surface of particles of a metal oxide, electrons are captured, and a resistive layer is formed due to the electron-depleted space charge region in n-type particles or a conductive layer from accumulated holes in p-type particles. Figure 11.1 shows the gas-sensing mechanism of the metal–oxide sensor. As the gas molecule interacts with chemisorbed oxygen, some of the electrons flow back to the metal oxide. A reaction would be $CO + O^- \rightarrow CO_2 + e^-$. The flow of those electrons changes the conductivity of the space charge layer; with n-type oxides, conductivity increases as the concentration of the charge-carrier electrons is greater; in the case of p-type oxides, conductivity decreases as electrons recombine with holes. This variation in conductivity within the space charge layer directly affects the overall electrical resistance of the oxide material. A thicker space charge layer leads to better sensitivity, and reducing particle size is beneficial for improving gas response since the space charge layer becomes

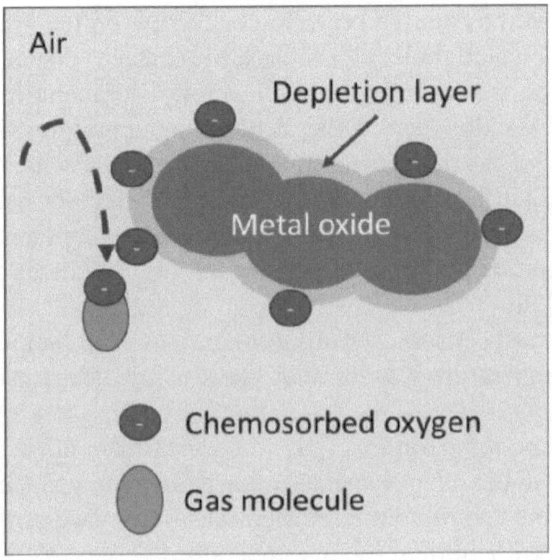

FIGURE 11.1 Gas-sensing mechanism of metal oxide sensor.

comparable to the volume of the gas-sensing oxide particle. Further, the concentration of charge carriers also is taken into consideration in terms of the thickness of the space charge layer; a higher concentration of the two semiconductors means a lower thickness of the layer since charge carriers easily reach the surface.

Divalent transition metal cations in spinel ferrites occupy octahedral sites with the potential for oxidation. Other such transitions include the oxidation of a divalent cation present at an octahedral site, such as Fe^{2+} in $ZnFe_2O_4$ or Ni^{2+} in $NiFe_2O_4$, to a trivalent state, Me^{3+}, which affects electrical conductivity. This mechanism agrees with the Verwey mechanism, where conductivity is achieved through electron hopping between the transition metal cations of the same element but different oxidation states, all found within the octahedral sites. The mechanism of electron hopping can thus be represented by the following equations: $Me^{3+} + e- \leftrightarrow Me^{2+}$ for n-type conductivity and $Me^{2+} + h^+ \leftrightarrow Me^{3+}$ for p-type conductivity. More availability of Me^{2+} and Me^{3+} pairs in the structure is equivalent to more cation conductivity. During the chemisorption of oxygen, the divalent cations are oxidized to trivalent cations. This oxidation process is reversed while reacting with gas at the surface, thus increasing or decreasing conductivity in n-type or p-type spinel ferrites, respectively. As in other metal–oxide–semiconductor gas sensors, n-type spinel ferrites show a decrease in resistivity after exposure to reducing gases, while the resistivity increases in p-type ferrites. Modification of the concentration of transition metal Me^{2+} cations in octahedral sites influences the amount of chemisorbed oxygen and the thickness of the space charge layer. Higher concentrations of Me^{2+} reduce the thickness of the space charge layer because the ease with which cations available at the surface of particles can be oxidized enhances the adsorption of chemisorbed oxygen.

11.3.1 Gas-Sensing Mechanism of n-Type Spinel Ferrites

The ideal stoichiometric composition of the n-type spinel ferrite gas sensors is rarely realized in practice. Instead, samples are often mixtures of Fe^{2+} and Fe^{3+} pairs, which occupy the octahedral sites before the chemisorption of oxygen. An important exception is magnetite, in which stoichiometric amounts of Fe^{2+} and Fe^{3+} exist on the octahedral sites, leading to n-type conductivity. However, magnetite is inappropriate for application in chemical gas sensors because of its inherent thermal instability. The nanoparticles of magnetite transform irreversibly at about 200°C, first to maghemite and then to hematite, a high-temperature polymorph of iron oxide. The hopping mechanism for electron transfer between the iron cations in the octahedral sites enhances carrier conduction in n-type spinel ferrites. A higher concentration of Fe^{2+} leads to increased electron conductivity by the reaction $Fe^{3+} + e^- \rightarrow Fe^{2+}$. When it comes into contact with oxygen that is chemisorbed onto the n-type spinel ferrite surface, Fe^{2+} remains oxidized at Fe^{3+}. The electrons that are released into the ferrite material reduce Fe^{3+} to Fe^{2+} when there is a reaction with oxygen that has come onto the surface and interacts with a target gas. This produces a higher Fe^{3+}/Fe^{2+} pair concentration, which enhances the overall electrical conductivity. The surface oxidation–reduction reactions of n-type spinel ferrites have been demonstrated to be reversible by various researchers.

Among the well-known n-type spinel ferrite gas sensors are $ZnFe_2O_4$, $CdFe_2O_4$, and $MgFe_2O_4$ with excellent applications due to their reversibility in electron exchange and great sensitivity to gases. Surface engineering techniques can be adapted to realize optimized n-type conductivity of spinel ferrites, particularly $Mg_{0.5}Zn_{0.5}Fe_2O_4$. A wet chemical route by an alumina template-assisted method has been used in the preparation of hollow tubes of $Mg_{0.5}Zn_{0.5}Fe_2O_4$ so that they get embedded in the alumina template (Figure 11.2). Although isolated tubes show p-type sensing behavior, the behavior is drastically different from that of the embedded tubes with the alumina template [37].

11.3.2 Gas-Sensing Mechanism of p-Type Spinel Ferrites

The p-type spinel ferrites are identified with the inverse spinel structure. In the inverse spinel structure, both the Me^{2+} transition metal cations and the Fe^{3+} iron ions occupy the octahedral positions. A very interesting example of this is $NiFe_2O_4$. It has been indicated that the p-type conductivity results from hole hopping between Ni^{2+} and Ni^{3+} at the octahedral site as given in the reaction: $Ni^{2+} + ^{+} \rightarrow Ni^{3+}$. $NiFe_2O_4$ contains Ni^{3+} ions because of lattice cation vacancies brought about by the excess oxygen that is attracted to the nickel ions during the synthesis process. As a means of charge compensation within the lattice, some of these nickel ions undergo oxidation and hence give rise to the presence of Ni^{3+}. Oxygen interstitials or cation vacancies constitute defects within the system of $NiFe_2O_4$ as a result of the incorporation of oxygen. These defects are compensated by the holes, leading to p-type conduction. Ordinarily, Ni^{3+} does not adopt a higher oxidation state in $NiFe_2O_4$. However, when the concentration of intrinsic Ni^{3+} is low and that of Ni^{2+} is relatively high, the chemisorption of oxygen is enhanced. The chemisorbed oxygen then reacts with the test gas. Consequently, this reaction changes the resistance. A greater concentration of chemisorbed oxygen enhances the response to the gas. This calls for a higher concentration of Ni^{2+} on the surface of $NiFe_2O_4$ [38]. Even with such a low concentration of 5 at% of Ni^{2+} having been substituted by Fe^{2+}, the corresponding p-type conductivity may be generated after sol–gel combustion [39]. Higher ratios of Fe^{2+} to Ni^{3+} also allow for greater electron mobility via electron transfer between $Fe^{3+} + e^- \rightarrow Fe^{2+}$.

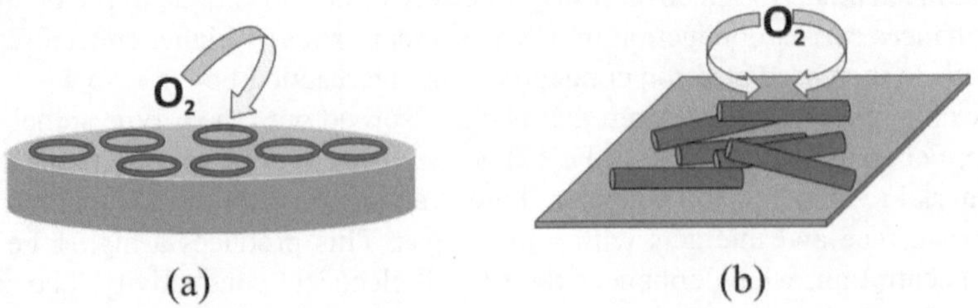

FIGURE 11.2 (a) Isolated nanotube and (b) embedded tube for gas sensing.

11.3.3 Advances in Ferrite-Based Gas Sensors

Recent advances have been shown by spinel ferrite gas sensors in that the inclusion of nanosized noble metal particles significantly enhances the gas response of semiconducting metal oxides. The sensitization can be through noble metals either electronically via silver (Ag) or palladium (Pd) or by chemical means using platinum (Pt) or gold (Au) [40–42]. Electronically sensitive species reduce the concentration of the space charge layer in the sensor and thus enhance the volume of gas response. On the other hand, chemical sensitizers increase the oxidation rate of the test gas at the surface but do not alter the resistance of the sensor. Noble metals increase the degree of gas response, increase selectivity, and decrease the operating temperature of the sensor. The finest results were obtained by Liu et al. They demonstrated that the inclusion of 1.5 wt% of Au significantly increased the response of $NiFe_2O_4$ to H_2S by an order of magnitude [44]. Meanwhile, Pt lowered the operating temperature of the $NiFe_2O_4$ sensor from its initial value of 300°C to 240°C [45]. Additionally, the inclusion of Pd nanoparticles within $MgFe_2O_4$ increased the gas response toward various gases, showing higher selectivity toward LPG as well as a lower operating temperature [46]. Similarly, the incorporation of 1.5 wt% Pd into $Ni_{0.6}Zn_{0.4}Fe_2O_4$ enhanced response and the response time for ethanol; it also gained selectivity toward ethanol and decreased the operating temperature by 70°C, reducing it from 300°C to 230°C [47].

Other carbon allotropes, such as carbon nanotubes or graphene, also present a gas response to various metal oxides, particularly SnO_2 [48], TiO_2 [49], WO_3 [50], ZnO [51], Fe_2O_3 [52], and NiO [53]. The carbon reduces the resistances of the sensing materials, thus increasing the electrical signals evoked from chemical signals [54]. It may then be concluded that no studies have been reported on the doping of carbonaceous allotropes into spinel ferrite gas sensors, which implies further study in this direction.

It is also reported that n-type and p-type semiconductor gas sensors can be hybridized to form p–n heterojunctions with better sensing performance. Some classifications of heterostructures are shown below:

i. Hybrid composite structures
ii. Bilayer and multilayer films

Structures carry with them second-phase particles, intermixed one-dimensional structures, and core–shell structures.

Recent advances in the field of spinel ferrite gas sensors have shown that doping with noble metal nanoparticles greatly enhances the response of gas from metal oxide semiconductors. Noble metals can sensitize spinel ferrites either electronically with Ag or Pd or chemically with Pt or Au [40–42]. Electronically sensitive species decrease the concentration of the space charge layer within the sensor, thereby enhancing the volume of gas response. In contrast, chemical sensitizers catalyze the oxidation of the test gas at the surface without affecting sensor resistance. Noble metals raise the degree of gas response, improve the selectivity, and lower the operating temperature of the sensor. The best results were achieved by Liu et al. Liu et al. demonstrated that the addition of 1.5 wt% Au enhanced the response of $NiFe_2O_4$ to H_2S by an order of magnitude [43–44]. Meanwhile, Pt reduced the operating temperature of the $NiFe_2O_4$ sensor down to 240°C from its original 300°C [45]. Also, Pd nanoparticles dispersed into $MgFe_2O_4$ improved the gas response toward various gases, including higher selectivity toward LPG, and reduced the operating temperature [46]. Likewise, the doping of 1.5 wt% Pd in $Ni_{0.6}Zn_{0.4}Fe_2O_4$ improved both the response and the response time for ethanol, developed selectivity toward ethanol, and reduced the operating temperature by 70°C from 300°C to 230 °C [47].

Other carbon allotropes like carbon nanotubes or graphene also provide gas response for other metal oxides, namely SnO_2 [48], TiO_2 [49], WO_3 [50], ZnO [51], Fe_2O_3 [52], and NiO [53]. Carbon reduces the resistances of the sensing materials, resulting in enhanced electrical signals from chemical signals [54]. It suggests that no one has ever doped carbonaceous allotropes into spinel ferrite gas sensors, and therefore, further research is still to be explored in this direction.

It has also been observed that n-type and p-type semiconductor gas sensors can be combined to form p–n heterojunctions, leading to further improvements in sensing performance [55]. Some classes of heterostructures are as follows:

(a) Mixed composite structures
(b) Bilayer and multilayer films
(c) Structures decorated with second-phase particles, mixed 1D–1D structures, and core–shell structures

11.4 CONCLUSION

In summary, ferrite gas sensors emphasize the versatility and effectiveness of ferrites as gas-sensing materials. The performance of these sensors is determined by their characteristics such as sensitivity, selectivity, operating temperature, and response time. These factors depend on preparation techniques, sintering conditions, and the specific gases detected. The gas-sensing properties vary for the respective ferrites. Doped ferrites always yield greater sensitivity than undoped ferrites. However, increases in sensitivity through doping often come with variations in response times as well as operating temperature. The aim here is to search for new materials that result in sensitive, selective, and durable devices for long-term use. As technology advances, the market for high-quality ferrite-based gas sensors will continue to grow. Scientific exploration of novel approaches will be crucial in meeting the growing demands for such sensors in a broad range of industrial applications, and thus contributing to better environmental observation and security.

REFERENCES

1. Yamazoe, N., & Miura, N. (1995). Development of gas sensors for environmental protection. *IEEE Transactions on Components, Packaging, and Manufacturing Technology: Part A, 18*(2), 252–256.
2. Punnoose, A., Reddy, K. M., Thurber, A., Hays, J., & Engelhard, M. H. (2007). Novel magnetic hydrogen sensing: A case study using antiferromagnetic haematite nanoparticles. *Nanotechnology, 18*(16), 165502.
3. Park, C. O., & Akbar, S. A. (2003). Ceramics for chemical sensing. *Journal of Materials Science, 38*, 4611–4637.
4. Pijolat, C., Pupier, C., Sauvan, M., Tournier, G., & Lalauze, R. (1999). Gas detection for automotive pollution control. *Sensors and Actuators B: Chemical, 59*(2–3), 195–202.
5. Dos Santos-Alves, J. S. G., & Patier, R. F. (1999). The environmental control of atmospheric pollution. The framework directive and its development. The new European approach. *Sensors and Actuators B: Chemical, 59*(2–3), 69–74.
6. Traversa, E. (1995). Design of ceramic materials for chemical sensors with intelligent properties. In: P. N Prasad, J. E. Mark & T. J. Fai, (Eds.), *Polymers and other advanced materials* (pp. 719–725). Springer.
7. Ramamoorthy, R., Dutta, P. K., & Akbar, S. A. (2003). Oxygen sensors: Materials, methods, designs and applications. *Journal of Materials Science, 38*, 4271–4282.
8. Barsan, N., & Schierbaum, K. (Eds.). (2018). *Gas sensors based on conducting metal oxides: Basic understanding, technology and applications*. Elsevier.
9. Tsang, S. C., & Bulpitt, C. (1998). Rare earth oxide sensors for ethanol analysis. *Sensors and Actuators B: Chemical, 52*(3), 226–235.
10. Huang, X. J., & Choi, Y. K. (2007). Chemical sensors based on nanostructured materials. *Sensors and Actuators B: Chemical, 122*(2), 659–671.
11. Kong, J., Franklin, N. R., Zhou, C., Chapline, M. G., Peng, S., Cho, K., & Dai, H. (2000). Nanotube molecular wires as chemical sensors. *Science, 287*(5453), 622–625.
12. Bruno, L., Pijolat, C., & Lalauze, R. (1994). Tin dioxide thin-film gas sensor prepared by chemical vapour deposition: Influence of grain size and thickness on the electrical properties. *Sensors and Actuators B: Chemical, 18*(1–3), 195–199.
13. Schedin, F., Geim, A. K., Morozov, S. V., Hill, E. W., Blake, P., Katsnelson, M. I., & Novoselov, K. S. (2007). Detection of individual gas molecules adsorbed on graphene. *Nature Materials, 6*(9), 652–655.
14. Llobet, E. (2013). Gas sensors using carbon nanomaterials: A review. *Sensors and Actuators B: Chemical, 179*, 32–45.
15. Lalauze, R., Pijolat, C., Vincent, S., & Bruno, L. (1992). High-sensitivity materials for gas detection. *Sensors and Actuators B: Chemical, 8*(3), 237–243.
16. Sakale, G., Knite, M., & Teteris, V. (2011). Polyisoprene-nanostructured carbon composite (PNCC) organic solvent vapour sensitivity and repeatability. *Sensors and Actuators A: Physical, 171*(1), 19–25.
17. Xiangfeng, C., Xingqin, L., & Guangyao, M. (1999). The catalytic effect of SmInO3 on the gas-sensing properties of CdIn2O4. *Materials Science and Engineering: B, 64*(1), 60–63.
18. Lange, U., Roznyatovskaya, N. V., & Mirsky, V. M. (2008). Conducting polymers in chemical sensors and arrays. *Analytica Chimica Acta, 614*(1), 1–26.
19. Martinelli, G., & Carotta, M. C. (1992). Sensitivity to reducing gas as a function of energy barrier in SnO2 thick-film gas sensor. *Sensors and Actuators B: Chemical, 7*(1–3), 717–720.
20. Seiyama, T., Kato, A., Fujiishi, K., & Nagatani, M. (1962). A new detector for gaseous components using semiconductive thin films. *Analytical Chemistry, 34*(11), 1502–1503.

21. Xiangfeng, C. (2003). Sulfide-sensing characteristics of MFe_2O_4 (M = Zn, Cd, Mg and Cu) thick film prepared by co-precipitation method. *Sensors and Actuators B: Chemical, 96*(3), 504–508.
22. Barsan, N., Koziej, D., & Weimar, U. (2007). Metal oxide-based gas sensor research: How to? *Sensors and Actuators B: Chemical, 121*(1), 18–35.
23. Yamazoe, N. (1991). New approaches for improving semiconductor gas sensors. *Sensors and actuators B: Chemical, 5*(1–4), 7–19.
24. Kim, H. J., & Lee, J. H. (2014). Highly sensitive and selective gas sensors using p-type oxide semiconductors: Overview. *Sensors and Actuators B: Chemical, 192*, 607–627.
25. Chiorino, A., Ghiotti, G., Prinetto, F., Carotta, M. C., Gallana, M., & Martinelli, G. (1999). Characterization of materials for gas sensors. Surface chemistry of SnO_2 and MoO_x–SnO_2 nano-sized powders and electrical responses of the related thick films. *Sensors and Actuators B: Chemical, 59*(2–3), 203–209.
26. Korotcenkov, G. (2007). Metal oxides for solid-state gas sensors: What determines our choice?. *Materials Science and Engineering: B, 139*(1), 1–23.
27. Sugimoto, M. (1999). The past, present, and future of ferrites. *Journal of the American Ceramic Society, 82*(2), 269–280.
28. Singh, V. R. (2005). *Smart sensors: Physics, technology and applications. Indian Journal of Pure and Applied Physics, 43*, 7–16.
29. Darshane, S., & Mulla, I. S. (2010). Influence of palladium on gas-sensing performance of magnesium ferrite nanoparticles. *Materials Chemistry and Physics, 119*(1–2), 319–323.
30. Bahadur, D. (1992). Current trends in applications of magnetic ceramic materials. *Bulletin of Materials Science, 15*(5), 431–439.
31. Srivastava, R., & Yadav, B. C. (2015). Nanostructured $ZnFe_2O_4$ thick film as room temperature liquefied petroleum gas sensor. *Journal of Experimental Nanoscience, 10*(9), 703–717.
32. Tianshu, Z., Hing, P., Jiancheng, Z., & Lingbing, K. (1999). Ethanol-sensing characteristics of cadmium ferrite prepared by chemical coprecipitation. *Materials Chemistry and Physics, 61*(3), 192–198.
33. Singh, S., Singh, A., Yadav, R. R., & Tandon, P. (2014). Growth of zinc ferrite aligned nanorods for liquefied petroleum gas sensing. *Materials Letters, 131*, 31–34.
34. Reddy, K. M., Satyanarayana, L., Manorama, S. V., & Misra, R. D. K. (2004). A comparative study of the gas sensing behavior of nanostructured nickel ferrite synthesized by hydrothermal and reverse micelle techniques. *Materials Research Bulletin, 39*(10), 1491–1498.
35. Zhang, G., Li, C., Cheng, F., & Chen, J. (2007). ZnFe2O4 tubes: Synthesis and application to gas sensors with high sensitivity and low-energy consumption. *Sensors and Actuators B: Chemical, 120*(2), 403–410.
36. Prinz, G. A. (1998). Magnetoelectronics. *Science, 282*(5394), 1660–1663.
37. Zhuiykov, S., Ono, T., Yamazoe, N., & Miura, N. (2002). High-temperature NO_x sensors using zirconia solid electrolyte and zinc-family oxide sensing electrode. *Solid State Ionics, 152*, 801–807.
38. Reddy, C. G., Manorama, S. V., & Rao, V. J. (1999). Semiconducting gas sensor for chlorine based on inverse spinel nickel ferrite. *Sensors and Actuators B: Chemical, 55*(1), 90–95.
39. Jiang, Y., Song, W., Xie, C., Wang, A., Zeng, D., & Hu, M. (2006). Electrical conductivity and gas sensitivity to VOCs of V-doped $ZnFe_2O_4$ nanoparticles. *Materials Letters, 60*(11), 1374–1378.

40. Li, Z., Lai, X., Wang, H., Mao, D., Xing, C., & Wang, D. (2009). General synthesis of homogeneous hollow core–shell ferrite microspheres. *The Journal of Physical Chemistry C, 113*(7), 2792–2797.
41. Patil, K. C., Manoharan, S. S., & Gajapathy, D. (2021). Preparation of high-density ferrites. In N. P. Cheremisinoff (Ed.), *Handbook of ceramics and composites* (pp. 469–493). CRC Press.
42. Gadkari, A. B., Shinde, T. J., & Vasambekar, P. N. (2009). Structural analysis of Y^{3+}-doped Mg–Cd ferrites prepared by oxalate co-precipitation method. *Materials Chemistry and Physics, 114*(2–3), 505–510.
43. Chu, X., Jiang, D., & Zheng, C. (2006). The gas-sensing properties of thick film sensors based on nano-$ZnFe_2O_4$ prepared by hydrothermal method. *Materials Science and Engineering: B, 129*(1–3), 150–153.
44. Zhihao, Y., & Lide, Z. (1998). Synthesis and structural characterization of capped $ZnFe_2O_4$ nanoparticles. *Materials Research Bulletin, 33*(11), 1587–1592.
45. Chu, X., Jiang, D., & Zheng, C. (2006). The gas-sensing properties of thick film sensors based on nano-$ZnFe_2O_4$ prepared by hydrothermal method. *Materials Science and Engineering: B, 129*(1–3), 150–153.
46. Dube, G. R., & Darshane, V. S. (1993). Decomposition of 1-octanol on the spinel system $Ga1{-_x}Fe_xCuMnO_4$. *Journal of Molecular Catalysis, 79*(1–3), 285–296.
47. Mukherjee, K., & Majumder, S. B. (2009). Analyses of response and recovery kinetics of zinc ferrite as hydrogen gas sensor. *Journal of Applied Physics, 106*(6).
48. Gadkari, A., Shinde, T., & Vasambekar, P. (2010). Influence of rare-earth ions on structural and magnetic properties of $CdFe_2O_4$ ferrites. *Rare Metals, 29*, 168–173.
49. Jeseentharani, V., George, M., Jeyaraj, B., Dayalan, A., & Nagaraja, K. S. (2013). Synthesis of metal ferrite (MFe_2O_4, M = Co, Cu, Mg, Ni, Zn) nanoparticles as humidity sensor materials. *Journal of Experimental Nanoscience, 8*(3), 358–370.
50. Wolf, S. A., Awschalom, D. D., Buhrman, R. A., Daughton, J. M., von Molnár, V. S., Roukes, M. L., ... Treger, D. M. (2001). Spintronics: A spin-based electronics vision for the future. *Science, 294*(5546), 1488–1495.
51. Darshane, S. L., Suryavanshi, S. S., & Mulla, I. S. (2009). Nanostructured nickel ferrite: A liquid petroleum gas sensor. *Ceramics International, 35*(5), 1793–1797.
52. Liu, X., Xu, Z., Liu, Y., & Shen, Y. (1998). A novel high performance ethanol gas sensor based on CdO Fe2O3 semiconducting materials. *Sensors and Actuators B: Chemical, 52*(3), 270–273.
53. Sutka, A., Pärna, R., Mezinskis, G., & Kisand, V. (2014). Effects of Co ion addition and annealing conditions on nickel ferrite gas response. *Sensors and Actuators B: Chemical, 192*, 173–180.
54. Morrison, S. R. (1987). Selectivity in semiconductor gas sensors. *Sensors and Actuators, 12*(4), 425–440.
55. Arshaka, K., Twomey, K., & Egan, D. (2002). A ceramic thick film humidity sensor based on MnZn ferrite. *Sensors, 2*(2), 50–61.

12 Conclusion and Future Scope

Nupur Aggarwal, Naveen Kumar, and Anu Kapoor

12.1 SUMMARY OF KEY FINDINGS

12.1.1 Synthesis and Characterization

The methods used to prepare spinel and inverse spinel ferrites include the sol–gel, co-precipitation, hydrothermal, and solid-state methods. Each of the methods outlined has merits and demerits concerning the batch size, generation of non-desired impurities, and cost, etc. Thus, the sol–gel method can produce polymers with high purity and good distribution of the particles within the matrices, albeit being on the high side in terms of cost and complexity [1]. The co-precipitation method is less expensive and more amenable to mass production, but it may be difficult to manage particle size [2]. Hydrothermal synthesis can enable precise control over the size and crystallinity of the particles, but it is limited because it necessitates conditions of elevated temperature and pressure [3]. Solid-state reactions are relatively simple and amenable to scale-up; however, these processes mostly lead to larger particles and contaminants [4].

12.1.2 Magnetic Properties

Spinel and inverse spinel ferrites exhibit magnetic properties ranging from superparamagnetism to ferrimagnetism and even ferromagnetism, which depend on the appropriate cation distribution between the tetrahedral sites A and the octahedral sites B within the crystal lattice [5, 6]. The inversion degree, representing the portion of divalent and trivalent cations in the A and B sites, respectively, has enormous effects on magnetic behavior. The possibility of their tuning using changes in the synthesis process and chemical composition enormously widens their application domain. The magnetic properties of spinel and inverse spinel ferrites are determined by temperature, pressure, conditions, methods of synthesis, atomic and electronic structures, composition, microstructure, spin–orbit coupling of magnetic ions, etc., as represented in Figure 12.1.

Specific examples include the coercive cobalt ferrite, $CoFe_2O_4$, exhibiting high coercivity and a moderate magnetization level suitable for applications requiring permanent magnetic materials such as magnetic data storage [7, 8]. On the other hand, $MnFe_2O_4$ has low coercivity but high magnetization and will find applications

DOI: 10.1201/9781032665658-12

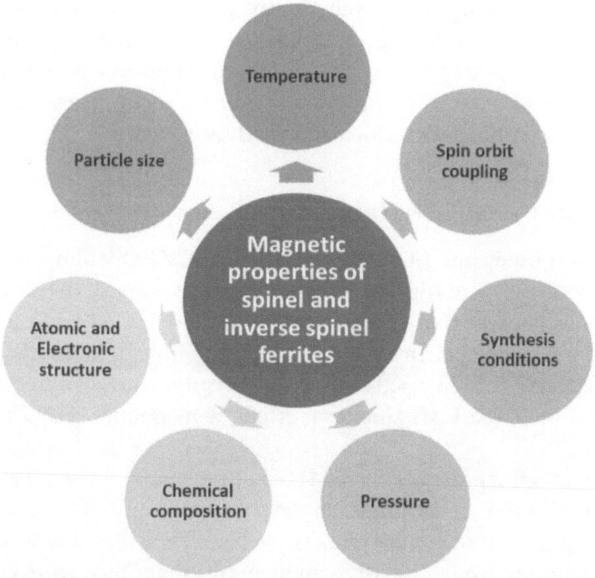

FIGURE 12.1 Dependence of magnetic properties of spinel and inverse spinel ferrites on various factors.

in uses like MRI contrasting agents. Magnetic properties can be tailored further by doping with other elements such as zinc, nickel, or copper to enhance desired traits in specific applications of interest.

12.2 APPLICATIONS OF SPINEL AND INVERSE SPINEL FERRITES

12.2.1 Magnetic Resonance Imaging (MRI)

Because of their high magnetic response and biocompatibility, ferrite nanoparticles look promising as contrast agents for MRI [9]. The more conventional gadolinium-based contrast agents are accepted to pose some risks of toxicity and nephrogenic systemic fibrosis in some cases, particularly in those where such products are used by patients with kidney-related problems. Ferrite nanoparticles would be a safer option with higher image contrast abilities.

Their use can significantly enhance the quality of the taken image, making the diagnostic images sharper and full of minute details. SPIONs (Superparamagnetic Iron Oxide Nanoparticles) have proved to be particularly effective in this respect because they generate strong local magnetic fields, which serve to shorten the relaxation times of hydrogen protons surrounding them and give higher contrast to MRI images [10]. Their surface functionalization with suitable biocompatible coatings, such as PEG (polyethylene glycol) or dextran, may further improve stability, circulation time, and targeting efficiency [11].

Ferrite-based contrast agents, in such a way, may allow for the clinical detection of tumors, inflammation, and other pathological states, as such materials will provide high-resolution pictures of soft tissues [12]. Current research is involved in the optimization of the size, surface properties, and magnetic characteristics of these nanoparticles for an enhanced performance and safety profile.

12.2.2 Electromagnetic Shielding

Spinel ferrites represent one of the most efficient EMI shielding materials due to their high magnetic permeability and electrical resistivity [13]. In this electronic age, EMI shielding has become obligatory to avoid interference seriously affecting sensitive equipment performance. Inverse spinel ferrites absorb and reflect electromagnetic waves and thus could be used in electronic equipment as an interfering barrier. Figure 12.2 illustrates the EMI shielding effect of magnetite (Fe_3O_4) nanoparticles [14].

Applications of spinel ferrites for EMI shielding spinel ferrites have been incorporated into composites or coatings applied on enclosures of electronic devices. These can be engineered to cover frequency usage from very low to very high and, therefore, will be effective in shielding applications that range from consumer electronics to aerospace systems. Besides, ferrites are lightweight and robust, hence the perfect choice for portable and high-performance applications.

Recent attempts have focused on the optimization of microstructure and the addition of conducting fillers, such as carbon nanotubes or graphene, to the ferrite composite for the enhancement of the shielding effectiveness of the ferrites. These hybrid materials will guarantee a better shielding performance while retaining the other valuable properties of the spinel ferrites.

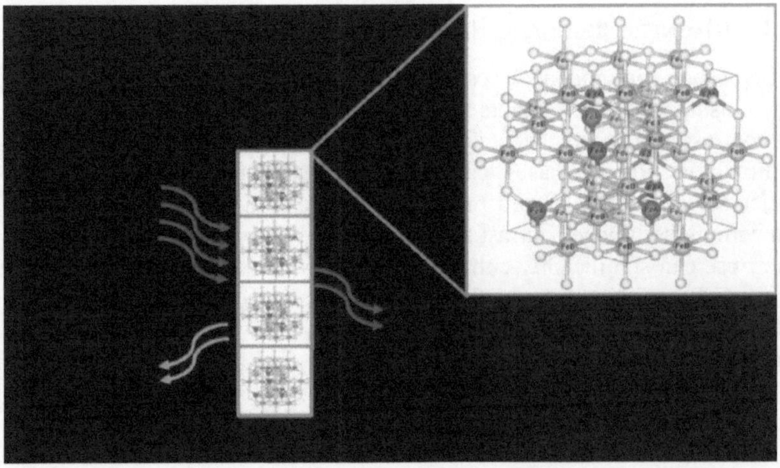

FIGURE 12.2 Illustration of the EMI shielding effect of magnetite (Fe_3O_4) NPs.

12.2.3 Drug Delivery

The application of spinel ferrites in drug delivery leverages their magnetic properties to target specific sites within the body, ensuring precise delivery of therapeutic agents [15]. Magnetic nanoparticles can be guided to the target site using an external magnetic field, minimizing side effects and improving treatment efficacy as shown in Figure 12.3. This targeted approach is particularly beneficial in cancer therapy, where high doses of chemotherapeutic drugs can cause significant harm to healthy tissues.

Spinel ferrites, such as iron oxide nanoparticles, are ideal candidates for magnetic drug delivery due to their biocompatibility and ease of functionalization. Surface modification with polymers, antibodies, or ligands allows for the attachment of drugs and targeting moieties, enabling specific interaction with cancer cells or other pathological tissues.

In addition to drug delivery, spinel ferrites are explored for their potential in hyperthermia treatment, where the nanoparticles generate localized heat upon exposure to an alternating magnetic field, causing thermal ablation of cancer cells [16]. Combining drug delivery and hyperthermia in a single platform could provide a synergistic effect, enhancing the overall therapeutic outcome.

12.2.4 Spintronics

Spinel ferrites possess great importance for the field of spintronics in general, where electron spin is used for storing and processing information [18]. In contrast with

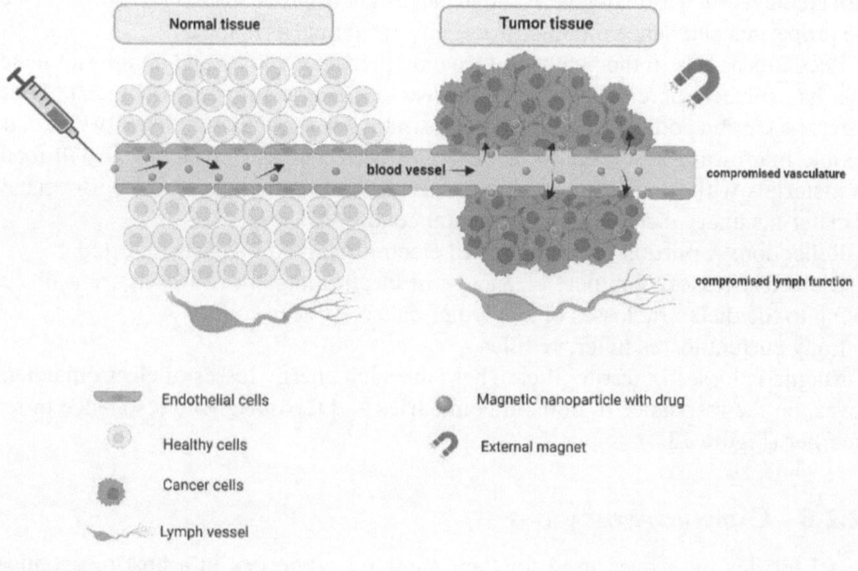

FIGURE 12.3 Illustration of the enhanced permeability and retention effect in cancer cells allows for passive targeting [17].

conventional electronics based on electron charge, spintronics takes advantage of the magnetic moment connected with the spin of an electron and thus opens new possibilities for constructing faster and energy-saving devices.

High spin polarization and low magnetic damping are essential to achieve effective spin injection and transport in spintronics. Spinel ferrites owe their high Curie temperatures and tunable magnetic properties to recently attracting much attention for spintronic applications. Magnetite (Fe_3O_4) is one of the most widely studied materials because its high spin polarization makes it into consideration in various applications in spin valves and magnetic tunnel junctions [19].

Spin-dependent properties in the case of spinel ferrites will be pursued in further studies concerning the mechanism of spin relaxation and interfacial effects, including other materials. Coupling spinel ferrites with magnetic or semiconducting materials into heterostructures and hybrid materials may provide new opportunities for advanced spintronic devices toward higher performance or even new functionalities.

12.2.5 Microwave Absorption

Spinel ferrites represent one of the most important active phases for microwave absorption materials due to their wide frequency range of electromagnetic radiation absorption [20]. Therefore, applications have been extended to stealth technology to reduce the radar cross-sectional area of military vehicles and also in electronic device protection by preventing electromagnetic interference. The influencing factors for the microwave absorption properties of spinel ferrites are particle size, shape, and composition. Optimization of these parameters will enable materials to achieve enhanced absorption efficiency with a broader bandwidth. Ferrites doped with elements such as manganese and nickel might improve the magnetic and dielectric properties and hence the microwave absorption performance.

Recent progress in the synthesis of nanostructured ferrites and composite materials has further increased their microwave absorption properties [21, 22]. These materials can be added to coatings or substrates for the making of lightweight and flexible microwave absorbers for a variety of uses. Further development will focus on materials with tunable absorption properties and on increasing the performance of materials under different environmental conditions.

Reflection: A portion of the energy of electromagnetic waves is reflected.

Dielectric losses: The part of energy in electromagnetic waves is transformed owing to the dielectric losses of a polymer matrix as heat.

Eddy current losses in ferrite filler.

Magnetic losses in ferrite filler: These are such energy losses of electromagnetic waves, on the resonance of domain boundaries and ferromagnetic resonance in ferrite filler (Figure 12.4).

12.2.6 Catalytic Applications

Spinel ferrites have been used for their catalytic properties in a host of chemical reactions, including oxidation and reduction [24]. Besides the fact that they can

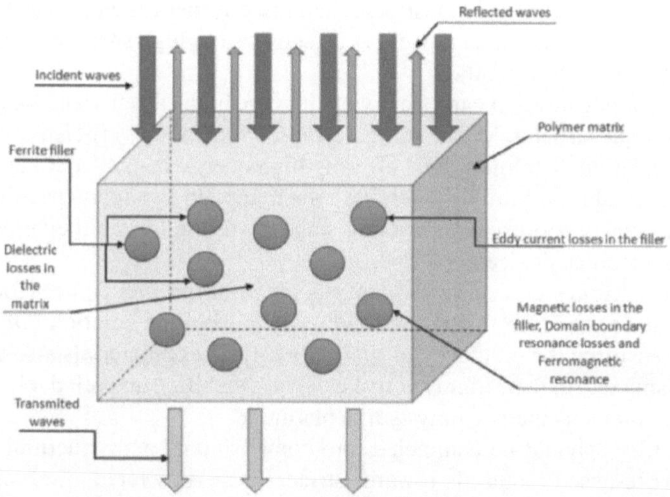

FIGURE 12.4 Scheme of the main mechanisms of absorption of electromagnetic waves in magnetic polymer composites with ferrite filler [23].

work under severe conditions, their stability and reusability make them attractive candidates for industrial applications of catalysts. Examples where ferrites are used include the catalytic decomposition of hydrogen peroxide, the oxidation of organic pollutants, and the reduction of nitrogen oxides in exhaust gases [25–27].

Certain factors that may affect the catalytic activity of ferrite spins include surface area, particle size, and chemical composition. Nanostructured ferrites, which feature high surface areas and specifically tailored surface properties, can exhibit higher catalytic performance. In addition, doping with transition metals or embedding them within composite materials can substantially lead to increased efficiency and selectivity in catalytic action.

Future research is directed toward the development of ferrite-based catalysts with improved performance and stability in various chemical reactions. Among those directions are (a) novel synthesis techniques to prepare nanostructured ferrites of precisely controlled properties and (b) the elaboration of hybrid materials that can combine the advantages of ferrites and other catalytic materials. Other promising areas of ferrite applications include environmentally friendly catalytic processes, like water splitting for hydrogen production and carbon dioxide reduction.

12.2.7 Energy Applications

Energy applications of spinel ferrites in battery electrodes, supercapacitors, and fuel cells have been intensively investigated. Their capability of enhancing energy storage/conversion efficiency is crucial for advancing sustainable energy technologies. For example, among the mentioned ferrites, nickel ferrite ($NiFe_2O_4$) [28] and cobalt ferrite ($CoFe_2O_4$) [29] have demonstrated their effectiveness as electrode materials in lithium-ion batteries and supercapacitors by showcasing high specific capacity, good conductivity, and stability [30].

In battery applications, it can effectively improve high capacity and long cycle life due to the unique structure of the spinel ferrites, which allows efficient intercalation and deintercalation of lithium ions, offering high energy density and fast charging/discharging capability. Similarly, ferrites, when applied in supercapacitors, exhibit high power density and excellent cycling stability, fitting for rapid energy delivery-based energy storage devices.

The spinel ferrites can serve as catalysts for both the fuel cell reactions of the oxygen reduction reaction (ORR) and the oxygen evolution reaction (OER), which are vital in ensuring the efficiency of fuel cells [31]. These materials also seem to be very promising due to their highly active catalysis, stability, and relatively lower cost compared to precious metal catalysts like platinum.

Future efforts should be channeled into compositional and structural optimizations of ferrite-based materials toward surface properties for high efficiency, stability, and scalability in energy applications. Besides, composite ferrites with other functional materials, such as graphene or carbon nanotubes, may achieve further performance improvements and extend their applications to advanced systems of energy storage and/or conversion.

12.3 FUTURE SCOPE

12.3.1 Advanced Synthesis Techniques

Advanced synthesis techniques will be developed in further studies to introduce properly controlled characteristics in both spinel and inverse spinel ferrites. Techniques such as microwave-assisted synthesis, atomic layer deposition, and electrospinning provide improved control over particle size, morphology, and cation distribution. These techniques can improve scalability and reduce the cost of ferrite production, hence making them more industrially accessible.

For example, microwave-assisted synthesis allows the generation of fast and homogeneous heating for a uniform particle size and enhancement of crystallinity. Atomic layer deposition allows the control of film thickness and composition at an atomic level, thus enabling the fabrication of high-quality thin films and nanostructures. Electrospinning helps in producing nanofibers with very high surface areas and controlled morphology and, therefore, is advantageous for applications requiring high catalytic activity or efficient ion transport.

Accordingly, the use of such advanced synthesis techniques can improve the performance of ferrites with desired properties. Moreover, a combination of different synthesis methods or even the incorporation of ferrites into other materials results in hybrid structures with synergistic effects, opening wider perspectives on their application.

12.3.2 Multifunctional Ferrites

The current research field holds promising advancements toward the development of multifunctional ferrites. These are ferrite materials that can simultaneously provide multiple desirable properties, such as high magnetization, catalytic activity, and electrical conductivity. Indeed, the use of multifunctional ferrites as a single piece may open the gateway to developing more efficient and versatile devices by abolishing the necessity for multiple components with less complex design and fabrication processes.

For example, the same ferrite material would be usable in both magnetic separation and catalytic degradation of pollutants because the same ferrite material would provide both high magnetic and catalytic performances. Ferrites with good electrical conductivities as well as magnetic properties could similarly find applications in advanced electronic devices and spintronic applications having superior performances and functionalities.

Based on these basic properties, current fundamental research in this area should be directed toward understanding the mechanism of multifunctional behavior inherent in ferrites and proposing strategies for optimizing their properties through synthesis control and doping. Furthermore, synergistic effects due to hybridizing ferrites with metals, oxides, or polymers are also worth exploring so that one may find new hybrid materials with improved performance and unprecedented functionalities.

12.3.3 Biomedical Applications

Further studies in terms of biocompatibility and toxicity should be conducted for extended adaptation of spinel ferrites in order to use them for biomedical applications. Though ferrites, particularly iron oxide nanoparticles, have potential applications in MRI contrast agents, drug delivery, and hyperthermia treatment, there should still be further in-depth research and testing on their long-term use as well as the possibility of side effects.

The study should be aimed at the synthesis of ferrites with higher biocompatibility, in addition to targeted functionality for hyperthermia treatment, bioimaging, and biosensing. This optimization is in terms of surface chemistry and the nature of the coating material so that such ferrite nanoparticles may have better biocompatibility, stability, and targeting efficiency. In vivo behavior, biodistribution, and clearance of the ferrite nanoparticles are important in order to ensure their safe and effective use in a clinical setting.

Multifunctional ferrite nanoparticles, through theranostic agents, are developed to possess both diagnostic functions and therapeutic functions. The simultaneous imaging and treatment agents would enable time-course monitoring of the responses toward the treatments and, thus, potentially improve their outcomes.

12.3.4 Environmental Applications

Environmental remediation might be one of the most impressive applications of spinel ferrites. The catalytic and adsorptive properties of these materials render them an effective tool for the removal of pollutants from air and water. Future studies should focus on the enhancement of such properties and the engineering cost-effective and scalable solutions for environmental cleanup.

An example application of ferrites as catalysts is in the degradation of organic pollutants, such as dyes, pesticides, and pharmaceuticals, by advanced oxidation processes. These ferrites have high catalytic activity, stability, and reusability – highly valued alternatives for conventional catalysts. Moreover, ferrites can be applied for the adsorption and removal of heavy metals and other toxic substances from contaminated water sources for sustainable water purification.

More studies are to be prompted into the generation of ferrite-based materials, with catalytic and adsorptive properties tailored through controlled synthesis methods, doping, and surface modification techniques. Further research efforts could go toward exploring the hybridization of ferrites with activated carbon, zeolites, or biochars for the development of hybrid materials with superior performance in environment-related applications.

12.3.5 Energy Storage and Conversion

The future of spinel ferrites in energy storage and conversion technologies for the use of lithium-ion batteries, supercapacitors, and fuel cells seems very promising. Efforts should be dedicated to improving their efficiency, capacity, and stability so that more competitive entities can be produced compared to the already existing technologies.

For instance, ferrite-based electrode materials exhibiting high specific capacity, good conductivity, and stability can contribute significantly to great improvements in battery performance. On the other hand, adjustments have to be made in ferrite compositions, structures, and morphologies to understand the improvement in their electrochemical properties. Another possible research area is finding new ferrite-based materials for solid-state batteries, which have enhanced safety characteristics along with high energy density.

Ferrite-based materials can provide high power density, excellent cycling stability, and rapid charge/discharge capabilities, which are efficient characteristics for energy storage with regard to applications that require very fast energy delivery in supercapacitors. Research focusing on the development of hybrid materials consisting of ferrites combined with other high-performance materials like graphene is expected to achieve superior electrochemical performance.

An extra new direction in fuel cells would be to further develop spinel ferrites as catalysts for ORR and OER in much more efficient energy conversion technologies at reduced costs. There is a continued focus on optimizing ferrite-based catalysts toward superior catalytic activity, stability, and scalability, followed by integration into advanced fuel cell systems.

12.3.6 Spintronics and Quantum Computing

The potential of spinel ferrites in spintronics and quantum computing is huge. Further research should be aimed at gaining insights and manipulating the spin-dependent properties of these materials so that their next-generation electronic device applications could potentially be faster and use much less energy. Spintronics exploits the magnetic moment connected with the spin of electrons, thus promising devices that are faster and use much less energy than classical electronics.

High spin polarization and low magnetic damping within the material are key to high performance in the injection and transport of spins in a spintronic system. Spinel ferrites look very promising for exploitation in spintronic applications, with a high Curie temperature and adjustable magnetic properties. For example, magnetite, Fe_3O_4, has been extensively studied because of its high spin polarization and therefore seems likely to become the basis of spin valves and magnetic tunnel junctions.

Other open questions and challenges remain in gaining an understanding of the detailed spin-dependent properties of ferrites, such as mechanisms of spin relaxation and interfacial effects with other materials. The development of heterostructures and hybrid material combinations of spinel ferrites with other magnetic or semiconducting materials may provide an impetus toward the development of new super-functionalized spintronic devices with enhanced performance and new functionalities.

Yet another interesting area of research is quantum computing, which employs qubits or quantum bits for information storage and processing. Materials exhibiting long spin coherence times and having strong spin–orbit coupling, like ferrites, are future candidates for use as qubit material, which will then enable the construction of ultra-reliable high-performance quantum computing devices.

12.3.7 Multifunctional and Hybrid Materials

Combination of properties of spinel ferrites with other functional materials such as polymers, metals, or oxides opens up the way for the development of advanced materials capable of improved performance and new functionalities based on multifunctional and hybrid materials. Hybrid material smay even take advantage of synergies from the combined benefits of the participating materials to display properties that are superior or more varied than those of the individual ones.

For example, the addition of spinel ferrites into polymer matrices could give rise to the creation of flexible yet lightweight composites that have improved magnetic, electrical, and mechanical properties. Thus, these materials can be used as a part of flexible electronics, sensors, or actuators to open new pathways for wearable and portable devices.

Hybrid materials with high electrical conductivity in combination with magnetic properties can be prepared by coupling spinel ferrites with a conductive material like graphene or carbon nanotubes. Such hybrid materials may therefore replace costly applications of electromagnetic shielding, apart from being used in advanced electronic devices and energy storage systems.

Following these directions, future research in this area should focus on new synthesis routes to fabricate hybrid materials with controlled composition, structure, and properties. It should also investigate interaction and interface phenomena that arise when different materials are combined in their hybrid structures to understand and optimize performance.

12.3.8 Isolators and Phase Shifters

Spinel ferrites have proven useful in microwave and radio frequency applications as isolators and phase shifters. An isolator is an element that allows signals to travel in just one direction but, in so doing denies reverse signal flow, which, on the other hand, protects sensitive components from damage or interference. Phase shifters, on the other hand, differ from isolators as they are designed to change the phases of signals. This is extremely important for beamforming and phase array antennas commonly used in communication and radar systems.

The magnetic properties of spinel ferrites, such as high saturation magnetization and low dielectric losses, serve to make them more suitable for these applications. Advances in synthesis and processing techniques have allowed for the development of ferrites with tailored magnetic and dielectric properties to further push the performance characteristics in isolators and phase shifters.

Future developments involve the miniaturization of spun ferrites in multifunctional devices; creating compact and efficient components for the next generation of communication systems should be pursued. Further work toward improving temperature stability, along with the application range, could help broaden the operational environment and applications of these materials.

12.3.9 Magnetic Data Storage

Spinel ferrites are suited to high-density magnetic data storage applications because of their relatively high coercivity and magnetic anisotropy. The magnetic stability of the materials forming the conventional magnetic storage media is essential to maintaining information stored within hard disk drives unchanged over time. They therefore appear to be a promising solution for future data storage technologies.

Information is encoded within magnetic data storage through the orientation of magnetic particles. The stability and switching behavior of the magnetic particles are very important in order to ensure reliable storage and retrieval of data. Ferrites like cobalt ferrite or $CoFe_2O_4$ have proven to be particularly appealing for such potential applications due to their high coercivity and thermal stability, thus assuring resistance to accidental data loss and permitting the storage of information at high densities.

Continuing research on ferrite-based materials is being directed toward developing new ferrite-based materials with better magnetic properties and compatibility with existing storage technologies. Part of this effort includes the development of nanostructured ferrites and composite materials combining the advantages of ferrites with other magnetic or conductive materials. Important progress is being made in lithography and deposition techniques for ferrite-based storage media with high density and nanoscale precision.

12.4 CONCLUSION

The applications of spinel and inverse spinel ferrites are as diverse as biomedical imaging, drug delivery, environmental remediation, and energy storage. These materials possess unique properties such as magnetic, catalytic, and electrical characteristics and are well-suited for advanced technologies. An integral part of the development of spinel ferrites, as research in this field continues, is the advancement of synthesis techniques, multifunctional materials, and hybrid structures. Future studies must focus on understanding the fundamental mechanisms underlying their properties and exploring new applications and integration strategies. Table 12.1 summarizes the key applications of spinel and inverse spinel ferrites.

The future of spinel and inverse spinel ferrites promises to lead the world toward innovative industries and the development of advanced technology. Researchers will have opportunities to open new avenues of challenges to be resolved through appropriate discussion and research into further unknown areas, paving the way toward innovative solutions and transformative advancements that will shape the science and future technological world. Thus, the continued research into the potential of spinel and inverse spinel ferrites can open many new possibilities for researchers and scientists working on modern solutions to problems that will arise in the future.

TABLE 12.1
Summary of Key Applications and Future Research Directions for Spinel and Inverse Spinel Ferrites

Application	Key Findings	Future Research Directions
Magnetic resonance imaging	Enhanced image contrast and safety	Improved biocompatibility and targeting
Electromagnetic shielding	High permeability and electrical resistivity	Hybrid materials with superior shielding performance
Drug delivery	Targeted delivery and hyperthermia treatment	Multifunctional theranostic agents
Magnetic data storage	High coercivity and stability	High-density storage and nanostructured materials
Spintronics	High spin polarization and tunable properties	Advanced spintronic devices and quantum computing
Microwave absorption	Effective radiation absorption	Tunable absorption properties and hybrid absorbers
Catalytic applications	High catalytic activity and stability	Environmentally friendly catalytic processes
Energy applications	High capacity and stability	Advanced electrode materials and fuel cell catalysts

REFERENCES

1. Zeshan, M., Waheed, M. S., Ishaq, S. A., Siddique, A., Ullah, A., Sherif, E. S. M., ... Farid, H. M. T. (2023). Sol–gel preparation of spinel ferrites/polyaniline composites for electromagnetic interface shielding. *Chemical Papers, 77*(11), 6533–6542.
2. Ren, P., Zhang, J., & Deng, H. (2009). Preparation and microstructure of spinel zinc ferrite $ZnFe_2O_4$ by Co-precipitation method. *Journal of Wuhan University of Technology-Mater. Sci. Ed., 24*(6), 927930.
3. Alqarni, A. N., Almessiere, M. A., Güner, S., Sertkol, M., Shirsath, S. E., Tashkandi, N., & Baykal, A. (2022). Structural and magnetic properties of hydrothermally synthesized Bi-substituted Ni–Co nanosized spinel ferrites. *Ceramics International, 48*(4), 5450–5458.
4. Cao, Y., Qin, H., Niu, X., & Jia, D. (2016). Simple solid-state chemical synthesis and gas-sensing properties of spinel ferrite materials with different morphologies. *Ceramics International, 42*(9), 10697–10703.
5. Hussein, M. M., Saafan, S. A., Abosheiasha, H. F., Zhou, D., Silibin, M. V., Trukhanov, S. V., ... Darwish, M. A. (2023). Impact of the Ni/Co ratio on structural and magnetic properties in A-site stoichiometric nanosized spinel ferrites. *Ceramics International, 49*(23), 39107–39116.
6. Darwish, M. A., Hussein, M. M., Saafan, S. A., Abd-Elaziem, W., Zhou, D., Silibin, M. V., ... Trukhanov, A. V. (2023). Impact of the Mg/Zn ratio on features of structural and magnetic properties in A-site stoichiometric nanosized spinel ferrites. *Journal of Alloys and Compounds, 968*, 172278.
7. Almessiere, M. A., Algarou, N. A., Slimani, Y., Sadaqat, A., Baykal, A., Manikandan, A., ... Ercan, I. (2022). Investigation of exchange coupling and microwave properties of hard/soft $(SrNi_{0\cdot02}Zr_{0\cdot01}Fe_{11\cdot96}O_{19})/(CoFe_2O_4)$ x nanocomposites. *Materials Today Nano, 18*, 100186.
8. Siva, K. V., Kumar, A., & Arockiarajan, A. (2021). Structural, magnetic and magneto-electric investigations on $CoFe_2O_4$ prepared via various wet chemical synthesis route: A Comparative Study. *Journal of Magnetism and Magnetic Materials, 535*, 168065.
9. Avasthi, A., Caro, C., Pozo–Torres, E., Leal, M. P., & García–Martín, M. L. (2020). Magnetic nanoparticles as MRI contrast agents. In A. R. Puente-Santiago & D. Rodríguez-Padrón (Eds.), *Surface-modified nanobiomaterials for electrochemical and biomedicine applications* (pp. 49–91). Springer.
10. Estelrich, J., Sánchez-Martín, M. J., & Busquets, M. A. (2015). Nanoparticles in magnetic resonance imaging: from simple to dual contrast agents. *International Journal of Nanomedicine*, 1727–1741.
11. Zhi, D., Yang, T., Yang, J., Fu, S., & Zhang, S. (2020). Targeting strategies for superparamagnetic iron oxide nanoparticles in cancer therapy. *Acta biomaterialia, 102*, 13–34.
12. Kefeni, K. K., Msagati, T. A., Nkambule, T. T., & Mamba, B. B. (2020). Spinel ferrite nanoparticles and nanocomposites for biomedical applications and their toxicity. *Materials Science and Engineering: C, 107*, 110314.
13. Yadav, R. S., Kuřitka, I., & Vilčáková, J. (2020). *Advanced spinel ferrite nanocomposites for electromagnetic interference shielding applications*. Elsevier.
14. Mokhosi, S. R., Mdlalose, W., Nhlapo, A., & Singh, M. (2022). Advances in the synthesis and application of magnetic ferrite nanoparticles for cancer therapy. *Pharmaceutics, 14*(5), 937.
15. Pusta, A., Tertis, M., Crăciunescu, I., Turcu, R., Mirel, S., & Cristea, C. (2023). Recent advances in the development of drug delivery applications of magnetic nanomaterials. *Pharmaceutics, 15*(7), 1872.

16. Patri, S., Thanh, N. T. K., & Kamaly, N. (2024). Magnetic iron oxide nanogels for combined hyperthermia and drug delivery. *Nanoscale, 16*, 15446–15464.
17. Mokhosi, S. R., Mdlalose, W., Nhlapo, A., & Singh, M. (2022). Advances in the synthesis and application of magnetic ferrite nanoparticles for cancer therapy. *Pharmaceutics, 14*, 937.
18. Rahman, M. M., Hasan, N., Tabassum, S., Rashid, M. H. O., Rashid, M. H., & Arifuzzaman, M. (2024). Bi^{3+} doped nanocrystalline Ni-Co-Zn spinel ferrites: Tuning of physical, electrical, dielectric and magnetic properties for advanced spintronics applications. *Ceramics International, 50*(21), 44585–44597.
19. Raut, S. D., & Dahotre, S. G. (2023). Synthesized cobalt substituted magnetite ($Co_xFe_{3-x}O_4$) nanoparticles–evaluation of structural, optical, and magnetic properties for spintronics application. *Ferroelectrics, 615*(1), 340–357.
20. Sahu, S., Mohapatra, P. P., Singh, H. K., & Dobbidi, P. (2023). Enhanced microwave absorbing performance of Sr^{2+} substituted Nickel-Cobalt nano ferrite for radar and stealth applications. *Materials Science and Engineering: B, 294*, 116514.
21. Sani, Y., Ismail, I., Yaakob, Y., & Mohammed, J. (2023). Enhanced electromagnetic microwave absorbing performance of carbon nanostructures for RAMs: A review. *Applied Surface Science Advances, 18*, 100455.
22. Mahmood, A., & Maqsood, A. (2023). High-frequency dielectric response of 3d metal (Mn and Cu) doped zinc ferrite nanoparticles for microwave applications. *Materials Today Communications, 34*, 105042.
23. Kostishin, V. G., Isaev, I. M., & Salogub, D. V. (2024). Radio-absorbing magnetic polymer composites based on spinel ferrites: A review. Polymers, *16*(7), 1003.
24. Anjaneyulu, B., Chinmay, Chauhan, V., Carabineiro, S. A., & Afshari, M. (2024). Recent advances on zinc ferrite and its derivatives as the forerunner of the nanomaterials in catalytic applications. *Journal of Inorganic and Organometallic Polymers and Materials, 34*(5), 1887–1907.
25. Starko, I., Tatarchuk, T., Naushad, M., & Danyliuk, N. (2024). Enhanced activity of la-substituted nickel–cobalt ferrites in congo red dye removal and hydrogen peroxide decomposition. *Water, Air, & Soil Pollution, 235*(8), 527.
26. Cheng, Y., Zhang, S., Wang, Z., Wang, B., You, J., Guo, R., & Zhang, H. (2023). Review on spinel ferrites-based materials (MFe_2O_4) as photo-Fenton catalysts for degradation of organic pollutants. *Separation and Purification Technology, 318*, 123971.
27. Ding, S., Li, C., Bian, C., Zhang, J., Xu, Y., & Qian, G. (2022). Application of low-cost MFe_2O_4 (M = Cu, Mn, and Zn) spinels in low-temperature selective catalytic reduction of nitrogen oxide. *Journal of Cleaner Production, 330*, 129825.
28. Mushtaq, M. W., Shahbaz, M., Naeem, R., Bashir, S., Sharif, S., Ali, K., & Dogar, N. A. (2024). Synthesis of surfactant-assisted nickel ferrite nanoparticles (NFNPs@ surfactant) to amplify their application as an advanced electrode material for high-performance supercapacitors. *RSC Advances, 14*(28), 20230–20239.
29. Das, S., Das, M. M., Dhamodharan, P., & Manoharan, C. (2022). Dielectric, magnetic and electrochemical characterization of $CoFe_2O_4$ nanoparticles and their application as an anode material in Li-ion battery. *Inorganic Chemistry Communications, 145*, 110050.
30. Malarvizhi, M., Meyvel, S., Sandhiya, M., Sathish, M., Dakshana, M., Sathya, P., ... Karthikeyan, S. (2021). Design and fabrication of cobalt and nickel ferrites based flexible electrodes for high-performance energy storage applications. *Inorganic Chemistry Communications, 123*, 108344.
31. Erdil, T., Ozgur, C., Geyikci, U., Lokcu, E., & Toparli, C. (2024). Earth-abundant divalent cation high-entropy spinel ferrites as bifunctional electrocatalysts for oxygen evolution and reduction reactions. *ACS Applied Energy Materials, 7*(18), 7775–7786.

Index